Skill: Counting money – penny

Henny Penny

Name ______________________

1 penny
1 cent

How much money?

Example

1¢ 1¢ = 5 ¢

1¢ 1¢ = ☐ ¢

1¢ 1¢ = ☐ ¢

1¢ 1¢ 1¢ = ☐ ¢

1¢ 1¢ 1¢
1¢ = ☐ ¢

1¢ 1¢ 1¢ 1¢
1¢ 1¢ 1¢ 1¢ = ☐ ¢

1¢ 1¢ 1¢ 1¢ 1¢ 1¢ 1¢ = ☐ ¢

Penny Penguin

Name ______________________________

How many cents?

Example

1¢ 1¢ 1¢ 1¢ = 4¢

1¢ 1¢ 1¢ 1¢ 1¢ 1¢ 1¢ 1¢ = ☐

1¢ 1¢ 1¢ 1¢ 1¢ = ☐

1¢ 1¢ 1¢ 1¢ 1¢ 1¢ 1¢ 1¢ 1¢ = ☐

1¢ 1¢ 1¢ = ☐

1¢ 1¢ 1¢ 1¢ 1¢ 1¢ = ☐

1¢ 1¢ 1¢ 1¢ 1¢ 1¢ 1¢ = ☐

1¢ 1¢ = ☐

1¢ 1¢ 1¢ 1¢ 1¢ 1¢ 1¢ 1¢ 1¢ 1¢ = ☐

Skill: Counting money – penny

You're Blooming Good!

Name ______________________________

Count the pennies on the flowers.
Write the cents in the center.

Penny Pinchers

Name ______________________________

Draw a line from the pennies to the right numbers.

Example

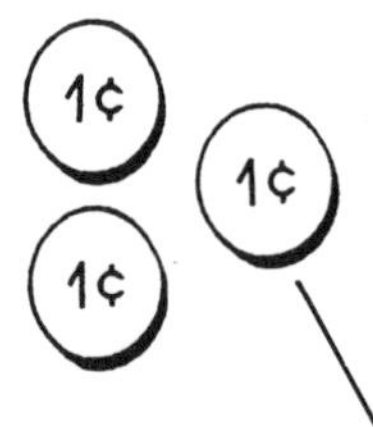

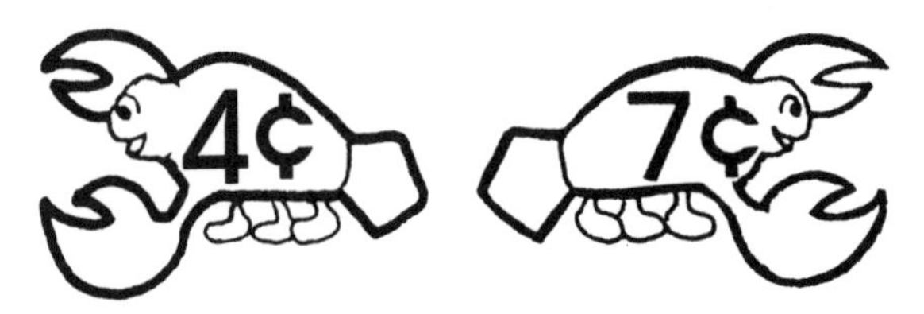

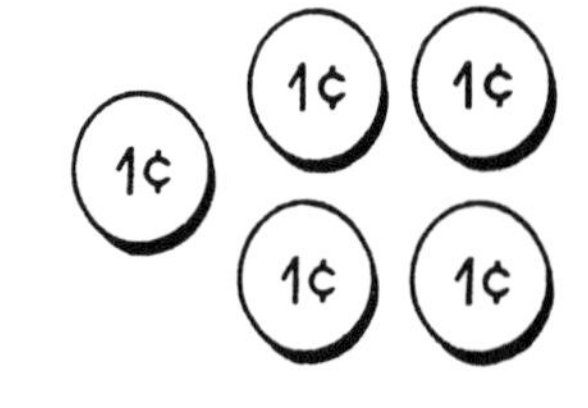

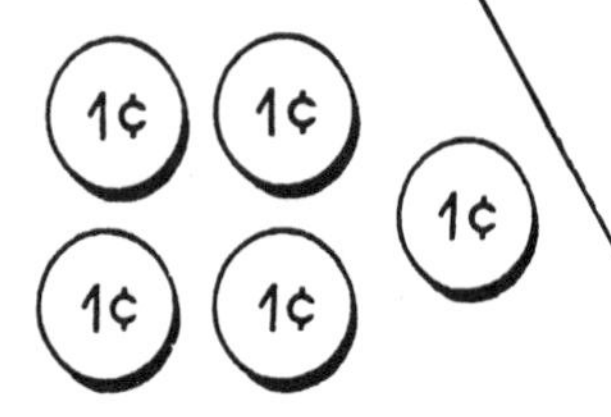

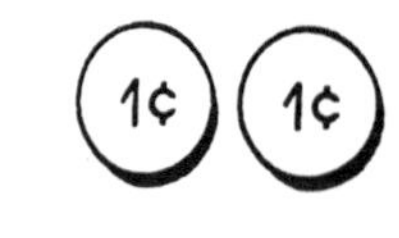

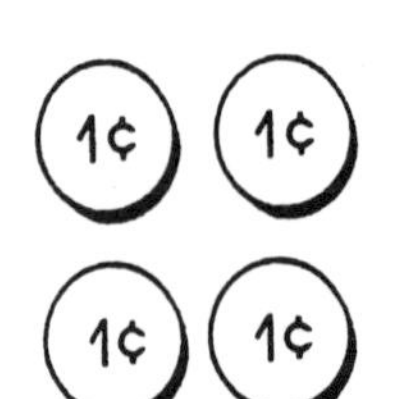

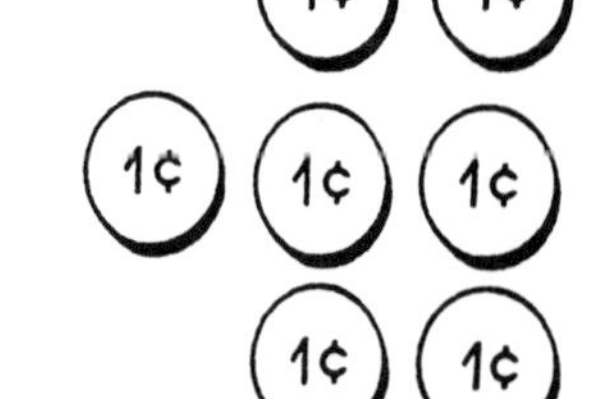

1¢ 1¢ 1¢
1¢ 1¢ 1¢
1¢ 1¢ 1¢

Nickel Pickles

Name ____________________

5 cents
1 nickel

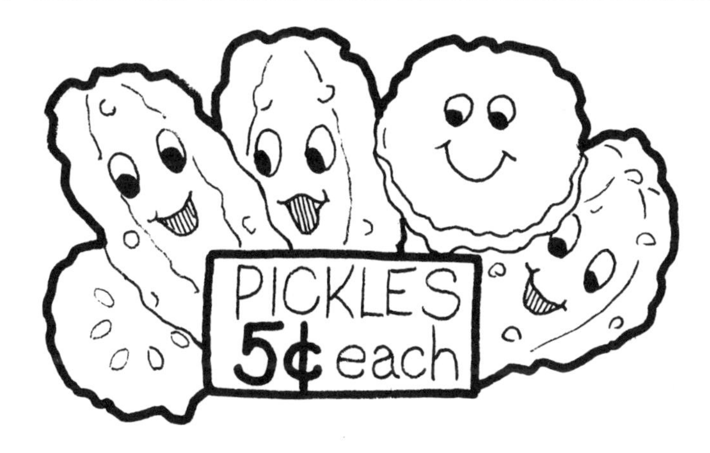

How much money?

Example

 = 15 ¢

Count 5 , 10 , 15

 = ☐ ¢

Count ___, ___

 = ☐ ¢

Count ___, ___, ___
___, ___

5¢ 5¢ 5¢ 5¢
5¢ 5¢ 5¢ = ☐ ¢

Count ___, ___, ___, ___
___, ___, ___

 = ☐ ¢

Count ___, ___, ___, ___

 = ☐ ¢

Count ___, ___, ___
___, ___, ___

Skill: Counting money – nickel

Clowns

Name ______________________________

Five Hive

Name ______________________

How much money is in each hive?

Example

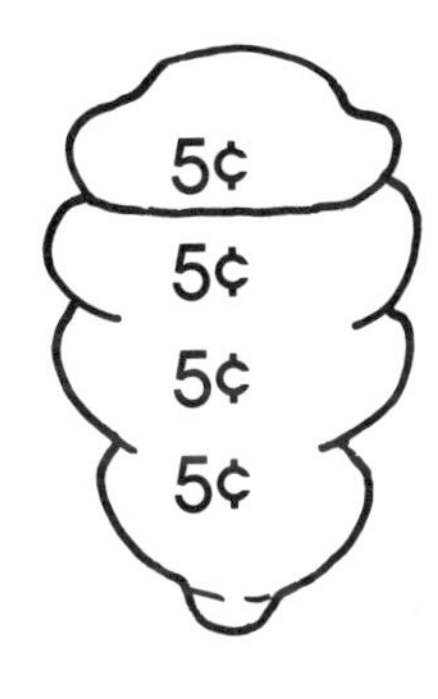

5¢ 5¢ 5¢ 5¢ — 20 ¢

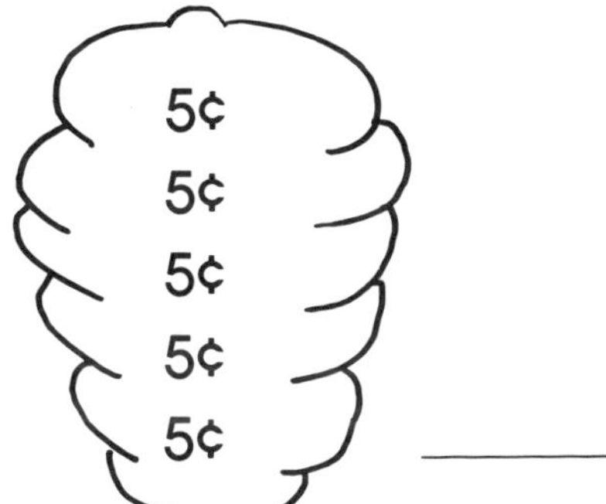

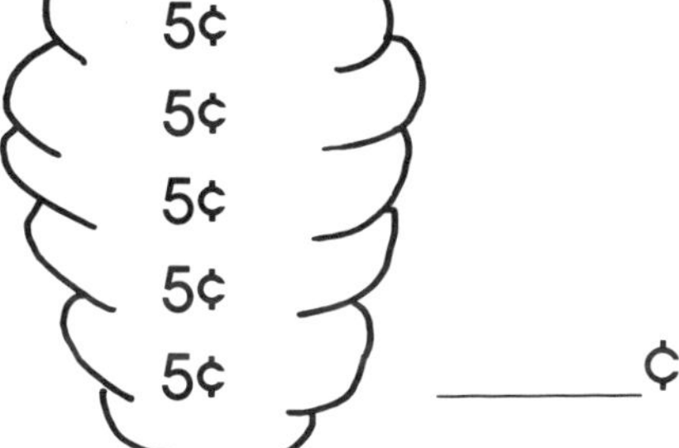

5¢ 5¢ 5¢ 5¢ 5¢ — ______ ¢

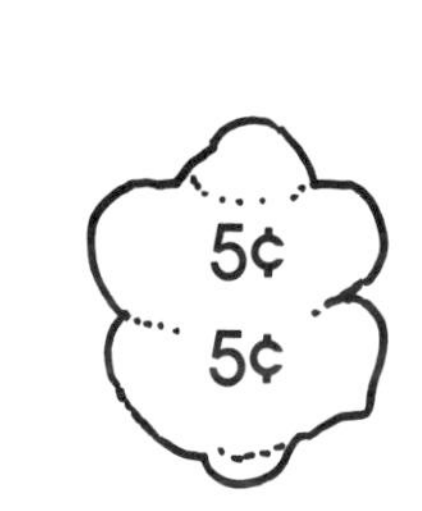

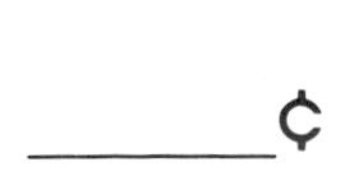

5¢ 5¢ — ______ ¢

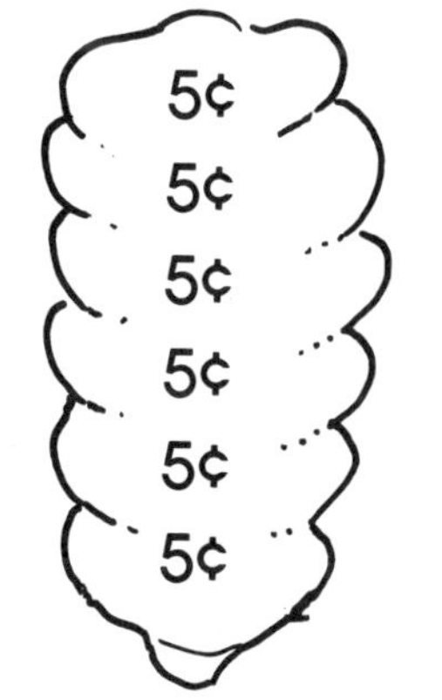

5¢ 5¢ 5¢ 5¢ 5¢ 5¢ — ______ ¢

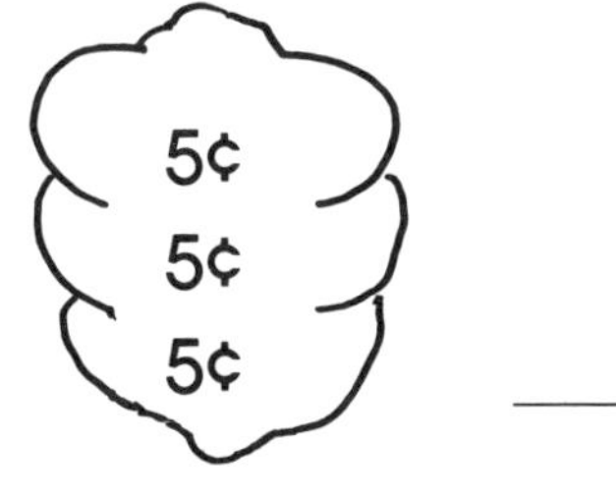

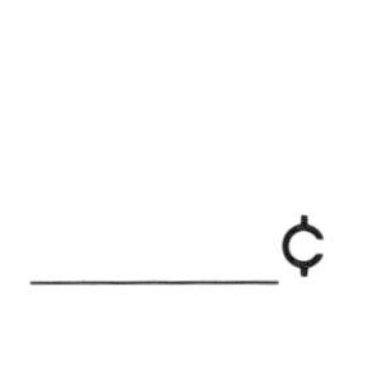

5¢ 5¢ 5¢ — ______ ¢

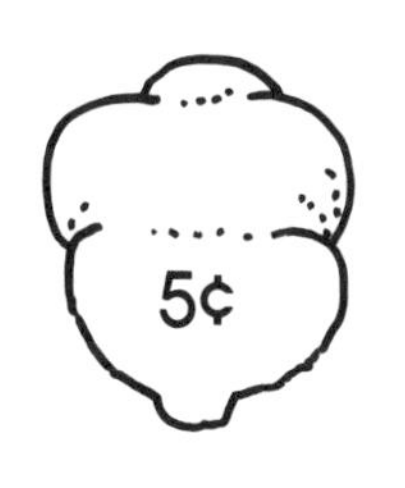

5¢ — ______ ¢

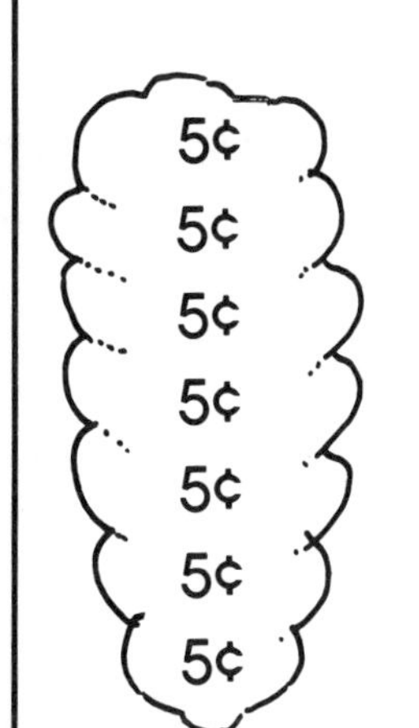

5¢ 5¢ 5¢ 5¢ 5¢ 5¢ 5¢ — ______ ¢

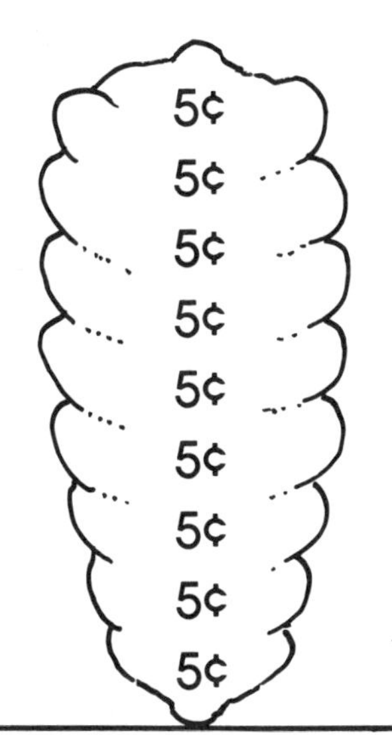

5¢ 5¢ 5¢ 5¢ 5¢ 5¢ 5¢ 5¢ 5¢ — ______ ¢

5¢ 5¢ 5¢ 5¢ 5¢ 5¢ 5¢ 5¢ — ______ ¢

Feed the Meter

Name ________________________________

Count the nickels. Write the money in the meter.

Example

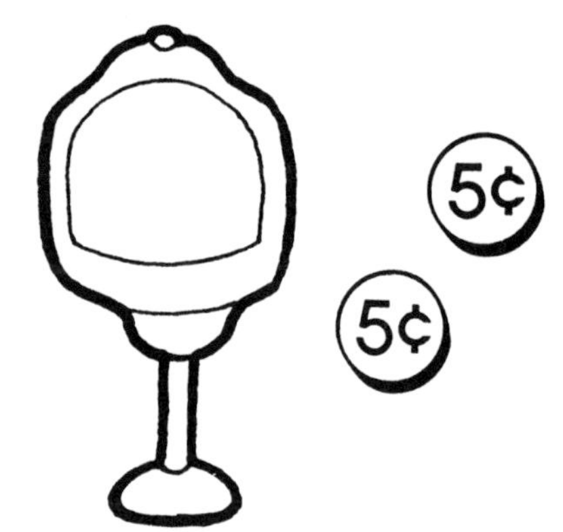

Skill: Counting money – penny, nickel

Money Bunnies

Name ______________________

Count the coins. Write the amount under each bunny's carrot.

Example

Skill: Counting money – penny, nickel

Hooting About Money

Name ______________________

Count the coins. Draw a line to match the owl with the same amount of money.

Example

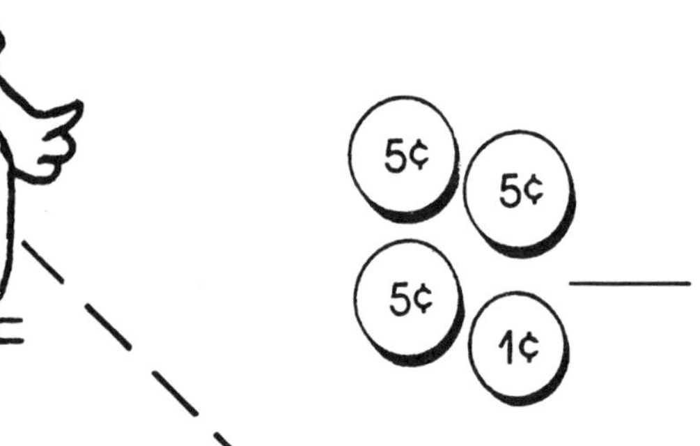

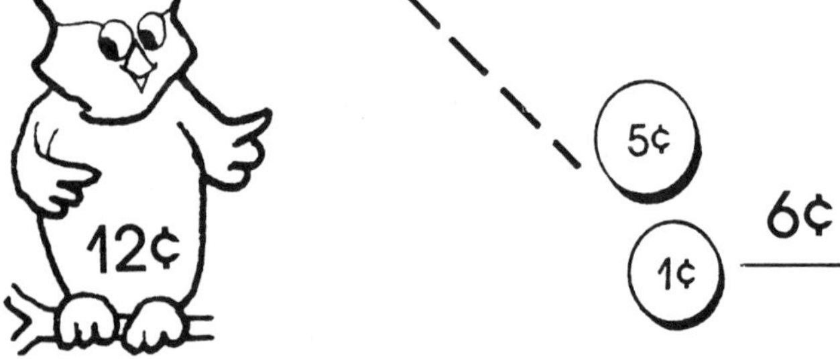

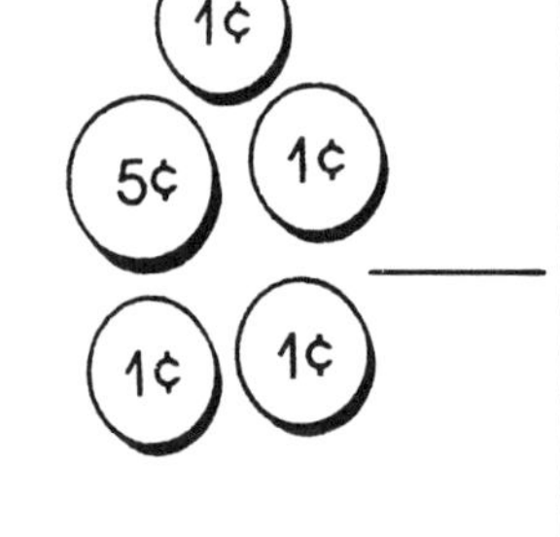

5¢ 5¢ 1¢ 1¢ ______

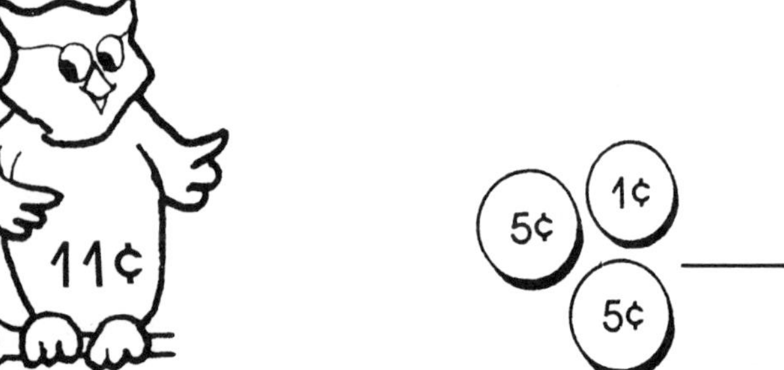

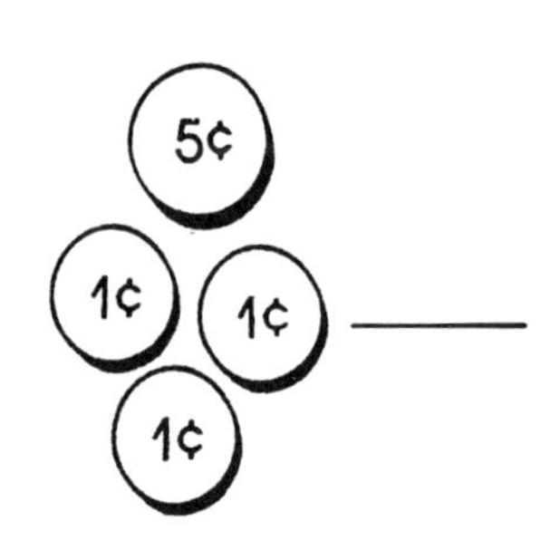

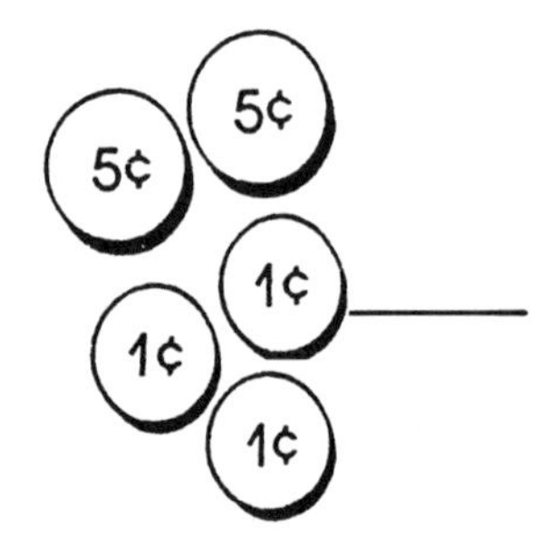

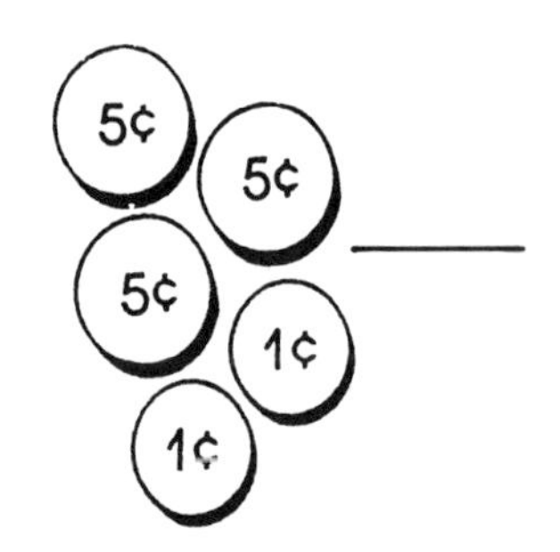

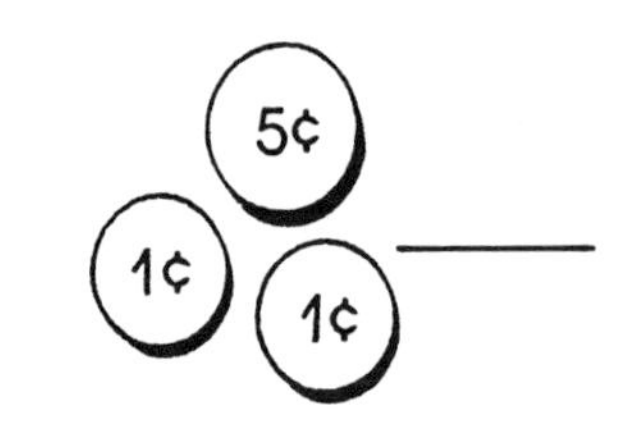

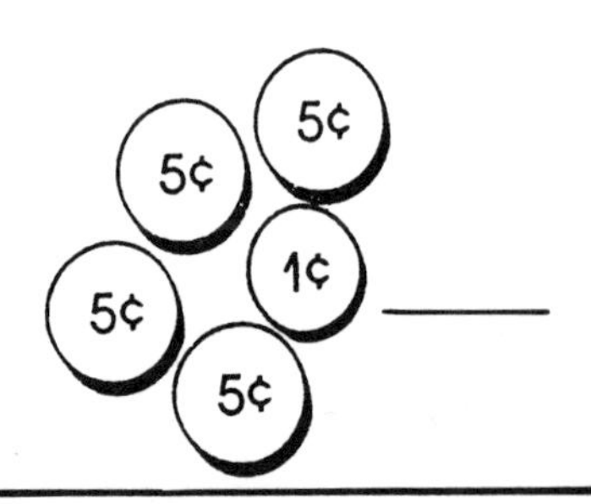

Cent — erpillars

Name ____________________

Count the coins on each "cent"erpillar

Example

17 ¢

_____ ¢

_____ ¢

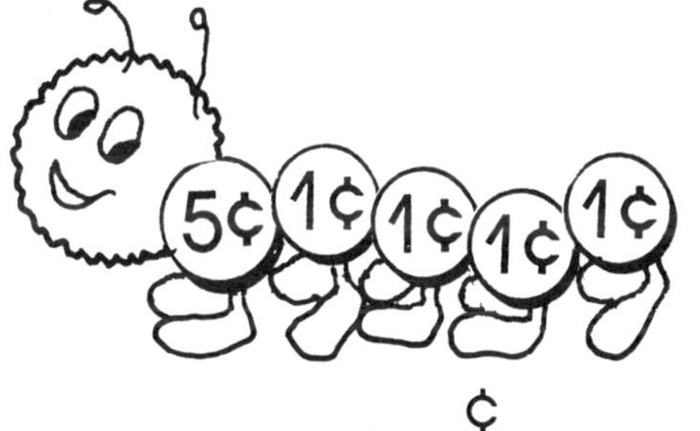

_____ ¢

_____ ¢

_____ ¢

5¢ 5¢ 5¢ 1¢ 1¢ 1¢

_____ ¢

_____ ¢

5¢ 5¢ 5¢ 5¢ 5¢ 5¢ 1¢

_____ ¢

5¢ 5¢ 1¢ 1¢ 1¢ 1¢ 1¢ 1¢ 1¢ 1¢ 1¢ 1¢

_____ ¢

Dime Climbs

Name ______________________________

Example

Climb the trees and count the money.
Write the answer under each tree.

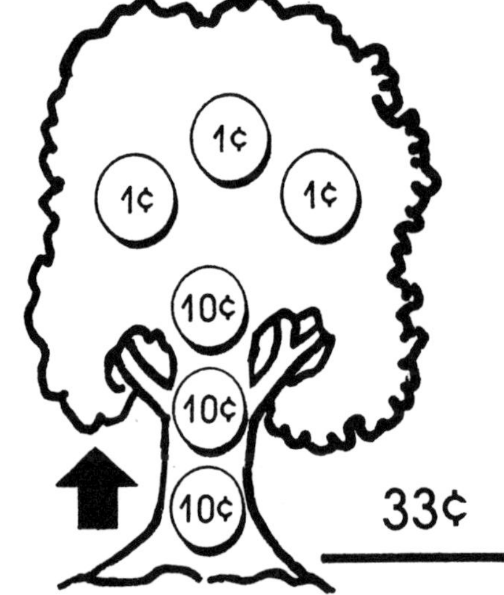

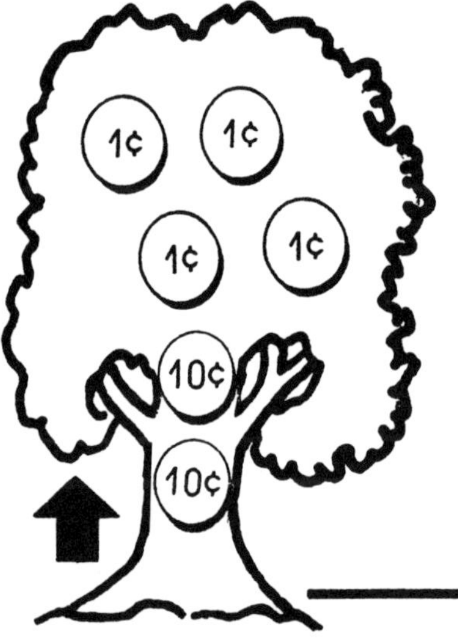

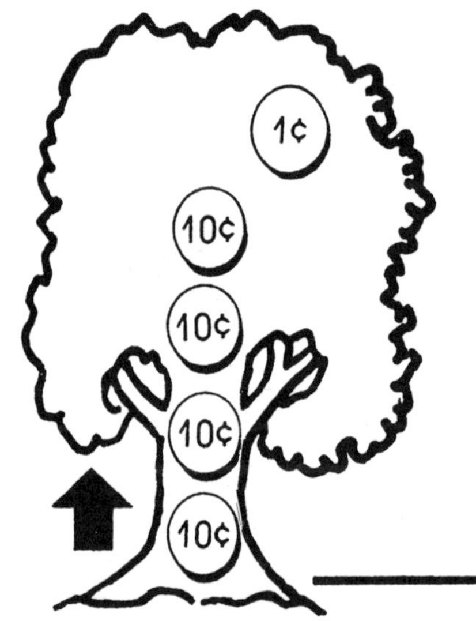

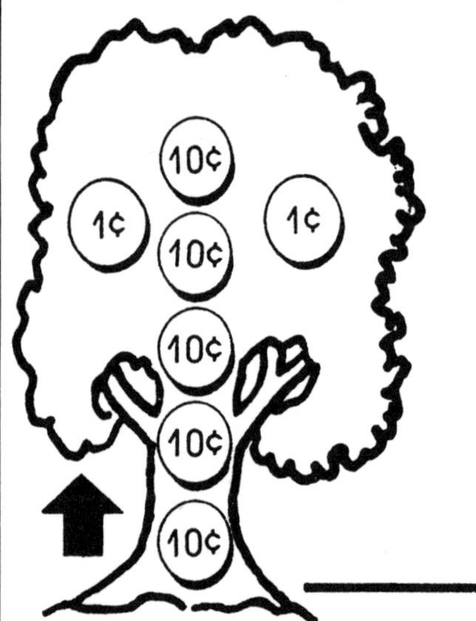

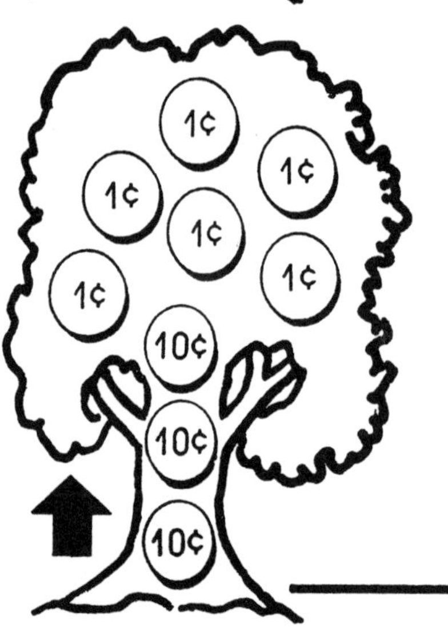

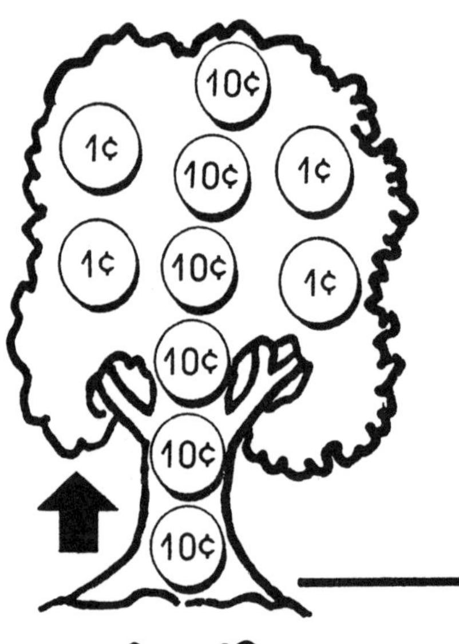

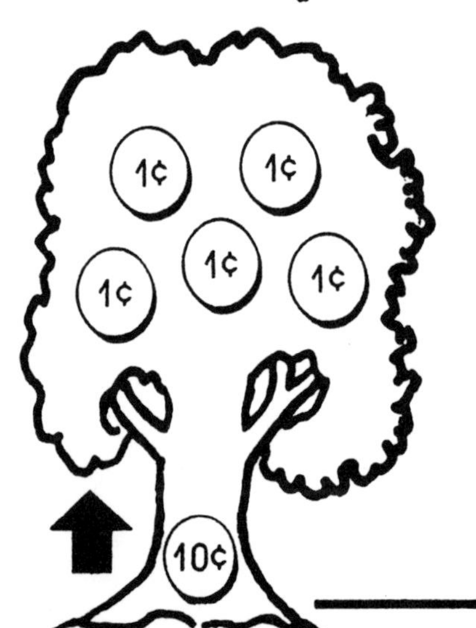

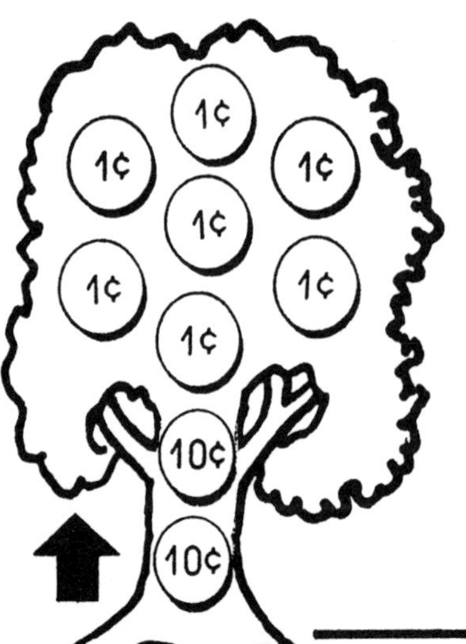

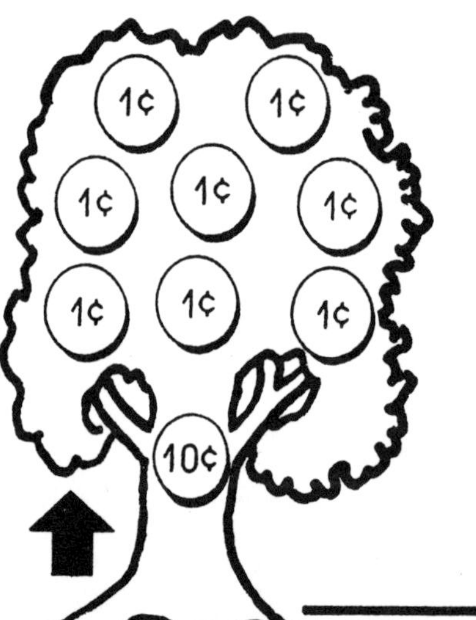

Skill: Counting exact change – dime, penny

Buy and Buy

Name ______________________________

Circle the coins to equal the right amount.

Example

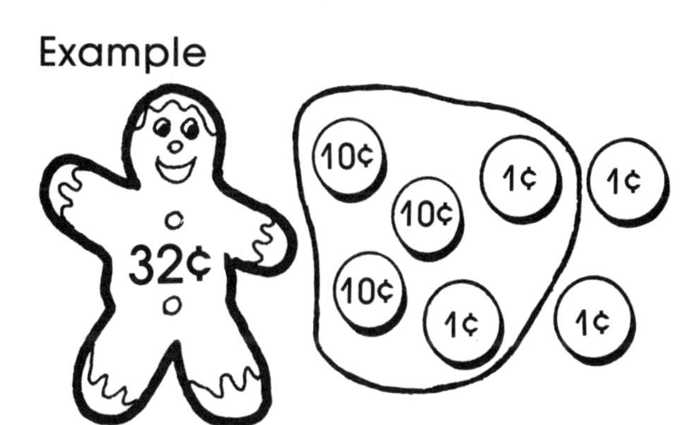

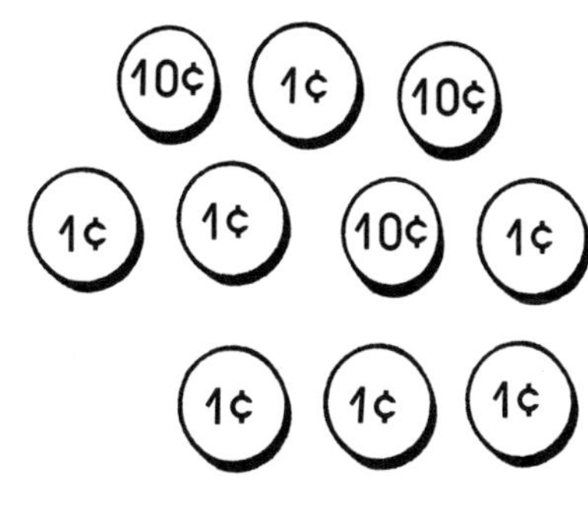

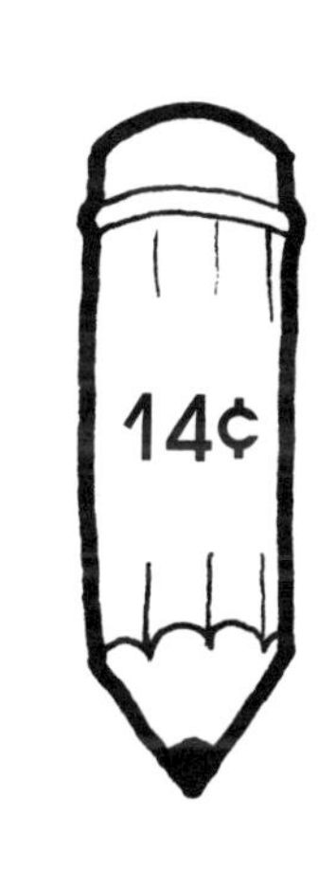

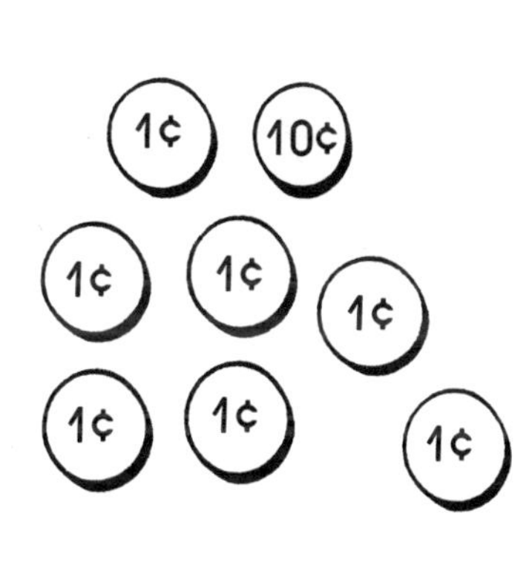

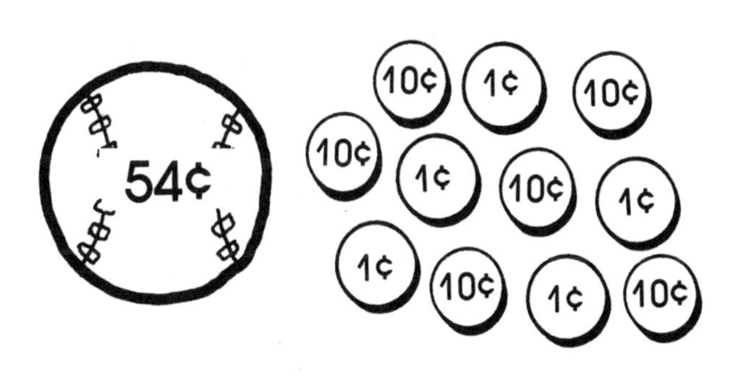

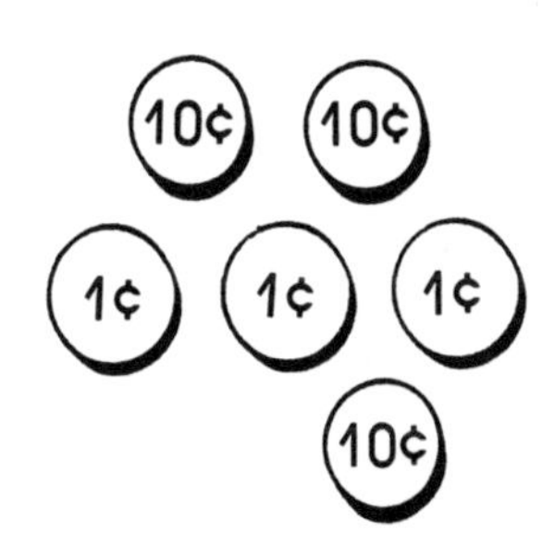

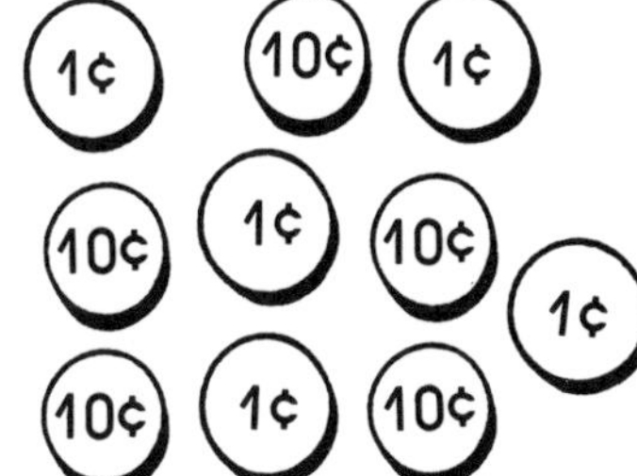

Money Belts

Name ______________________

Count the money on each belt. Write the amount under the belt.

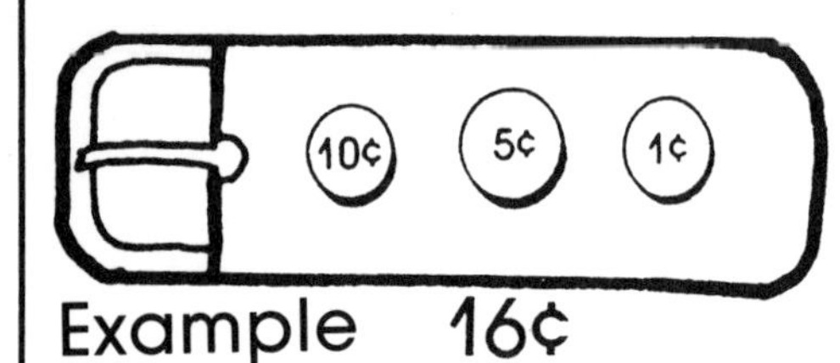

Example 16¢

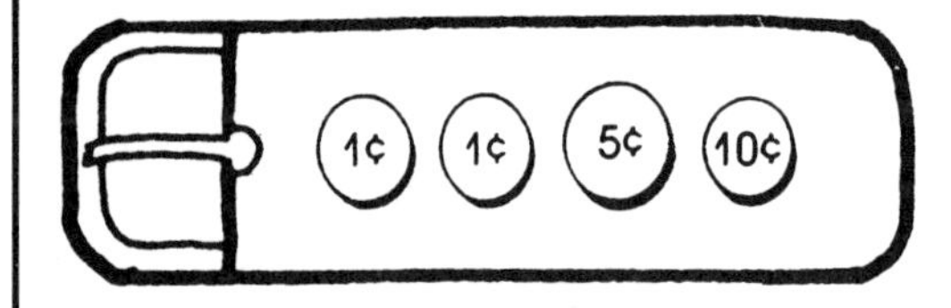

______

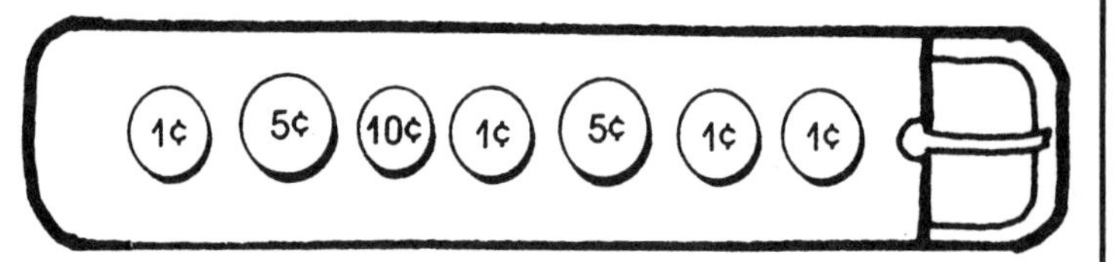

______

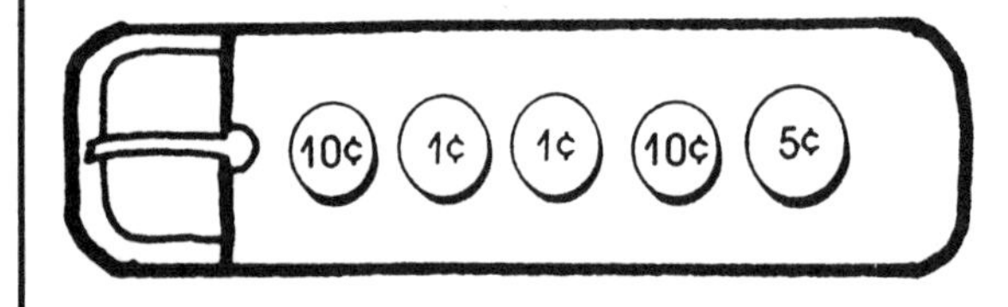

______

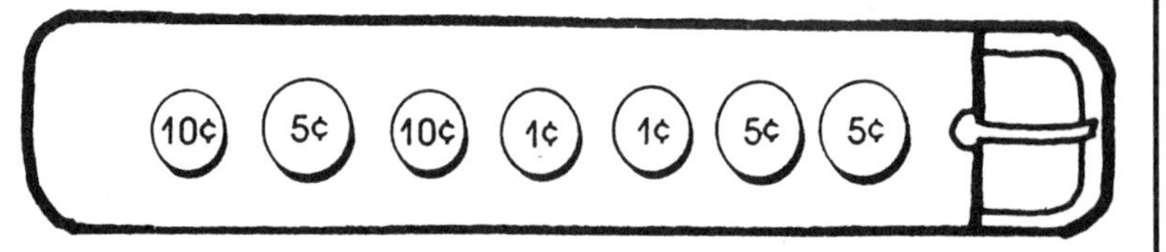

______

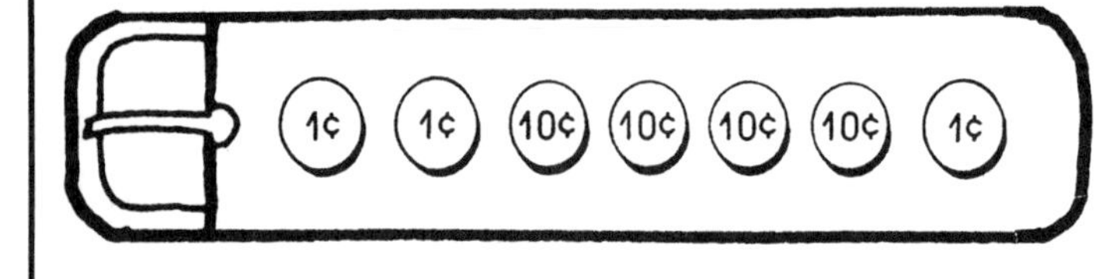

______

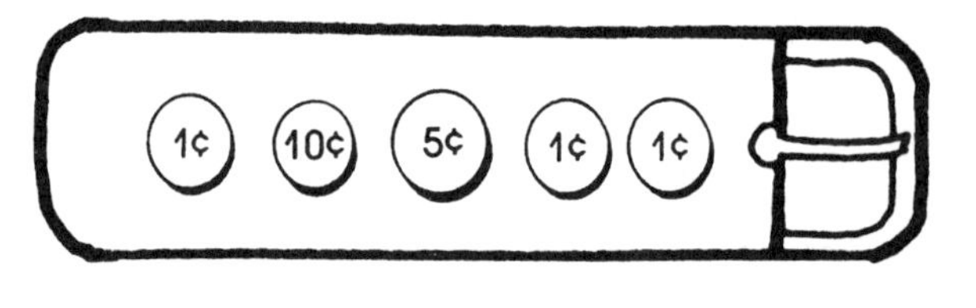

______

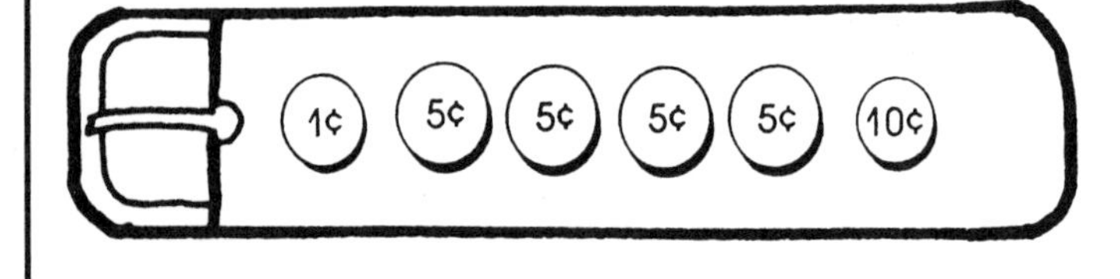

______

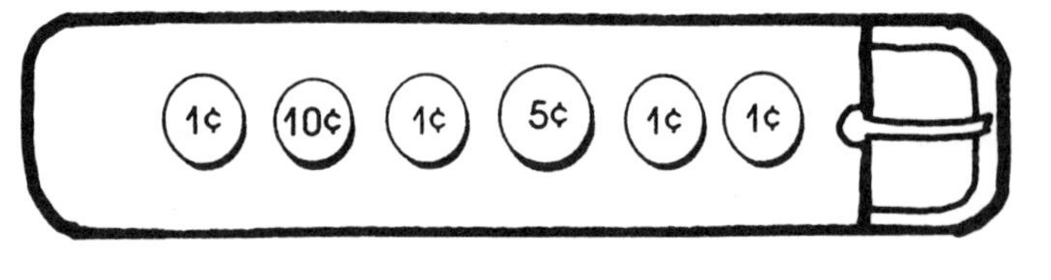

______

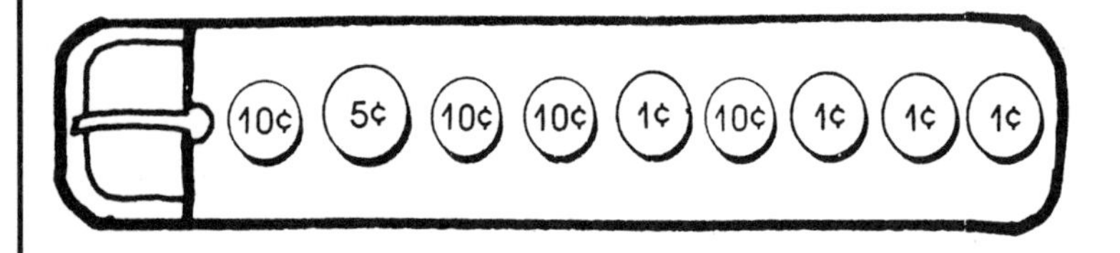

______

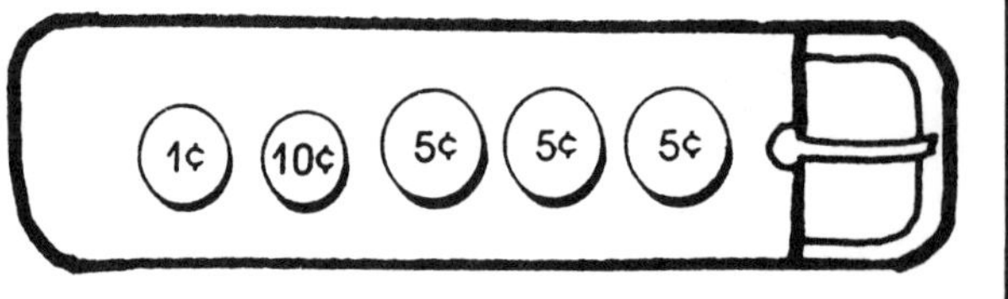

______

Skill: Money – comparing values (penny, nickel, dime)

"Sir Circle" Counts Coins!

Name ____________________

How much money?
Count the coins. Circle the set with more money.

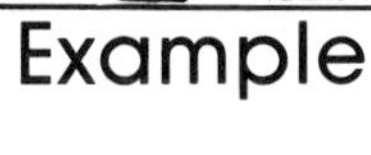

Example

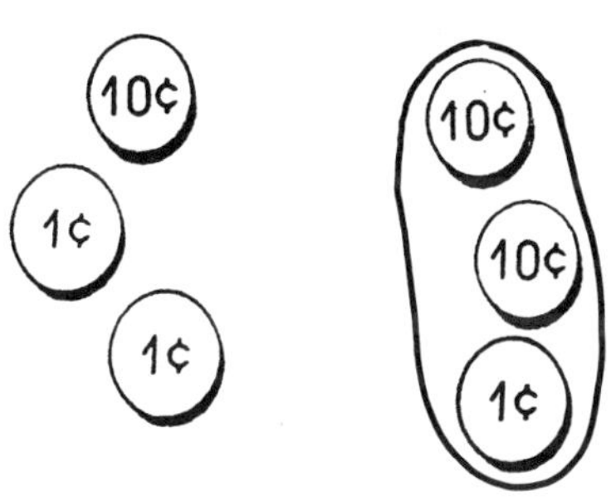

12¢ ______ 21¢ ______

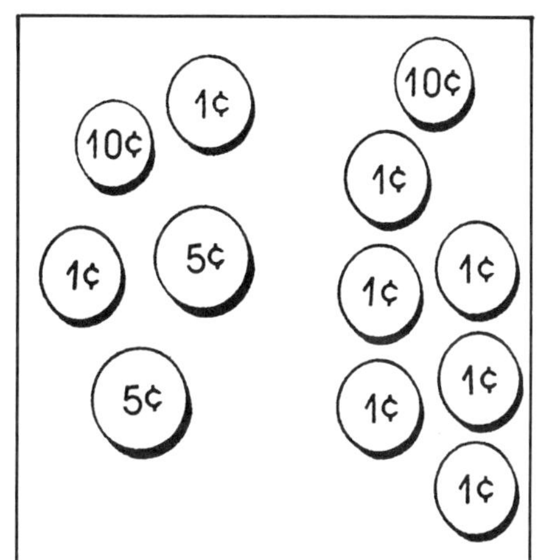

______ ______

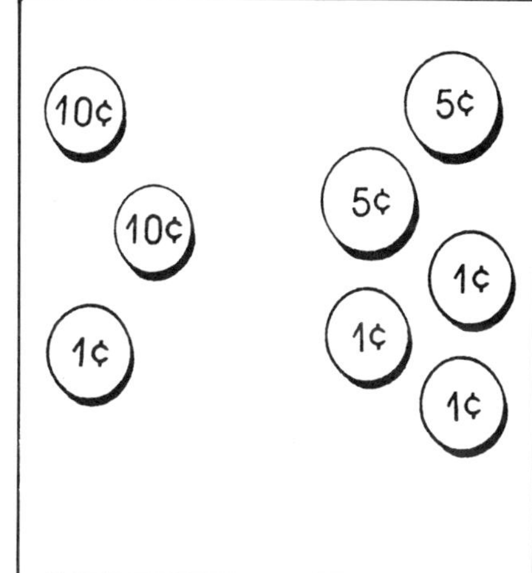

______ ______

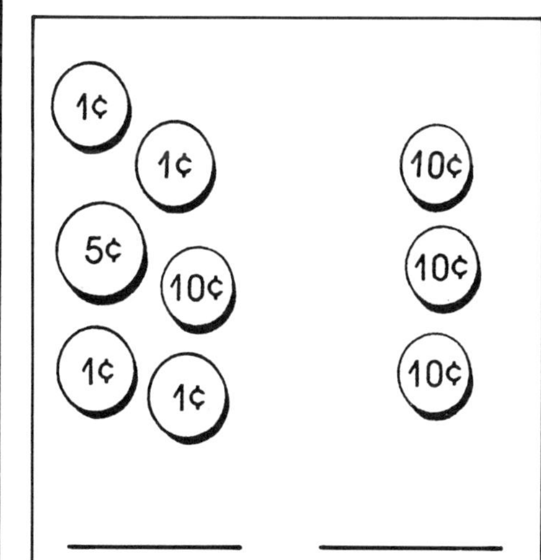

______ ______

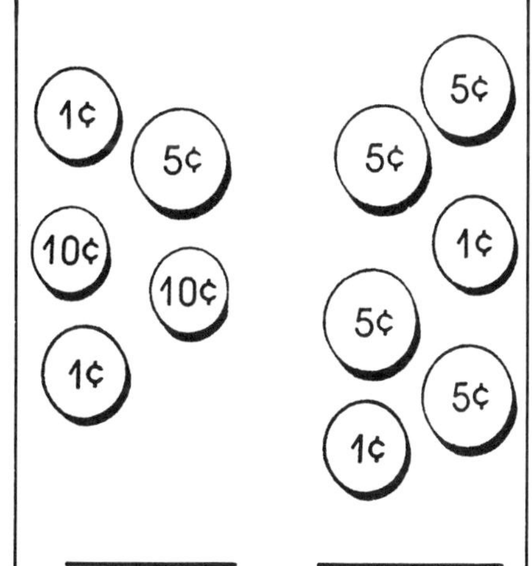

______ ______

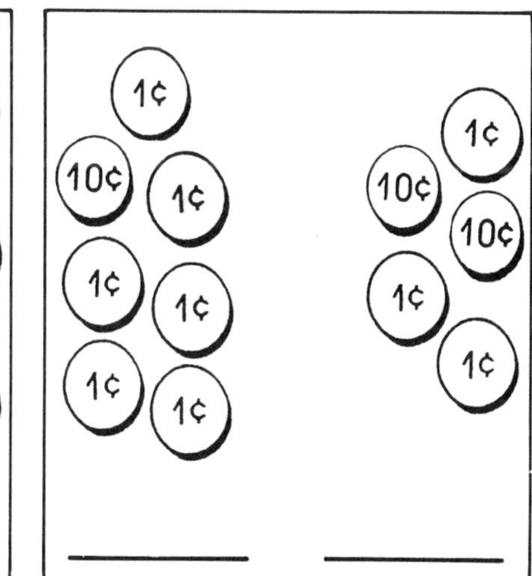

______ ______

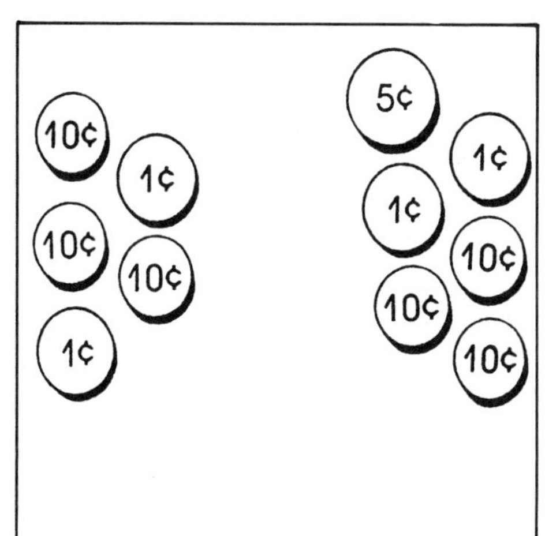

______ ______

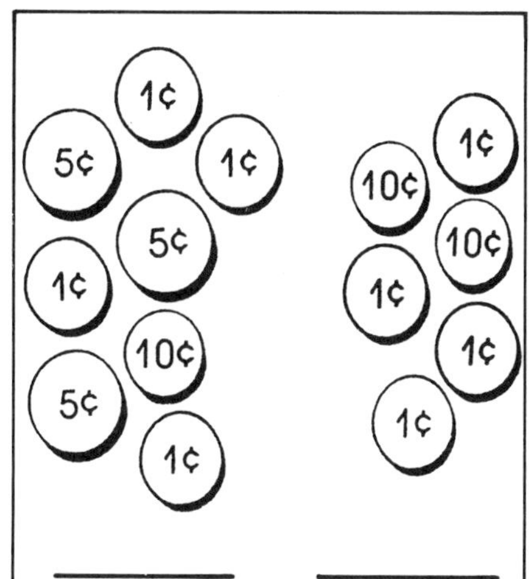

______ ______

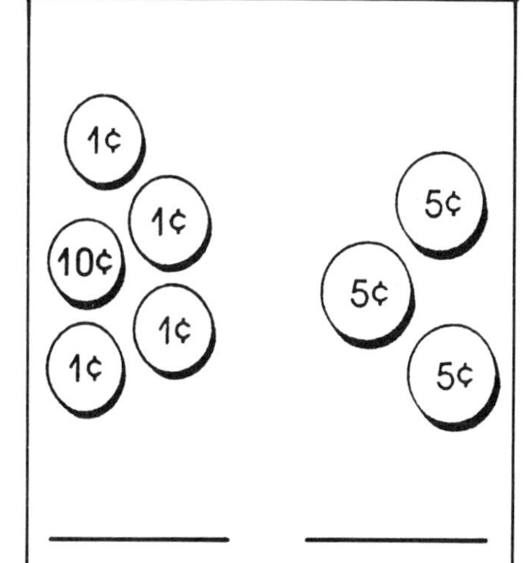

______ ______

Colorful Coins

Name ___________________________

Count the coins. Use the right color to circle the set of coins with the most money.

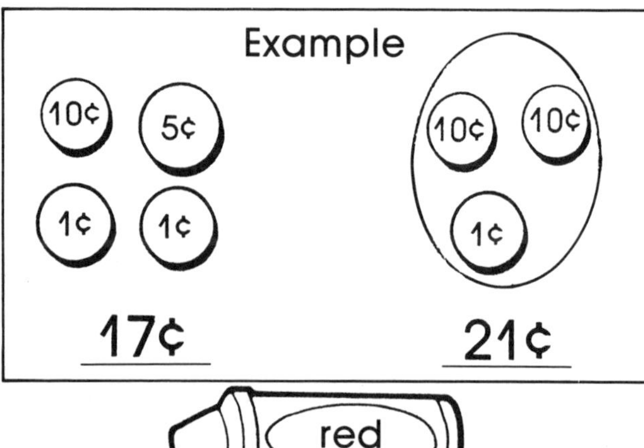

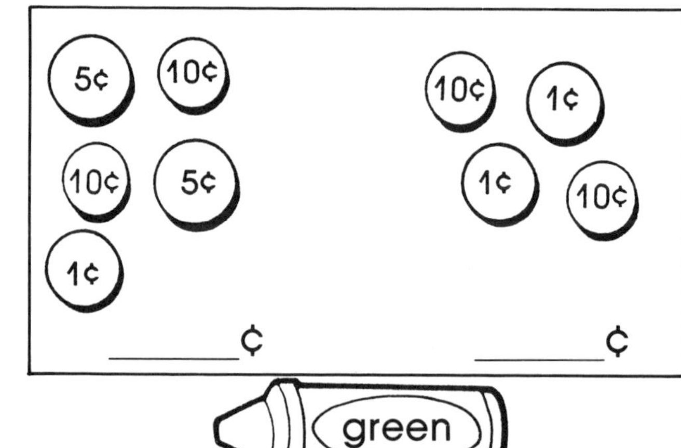

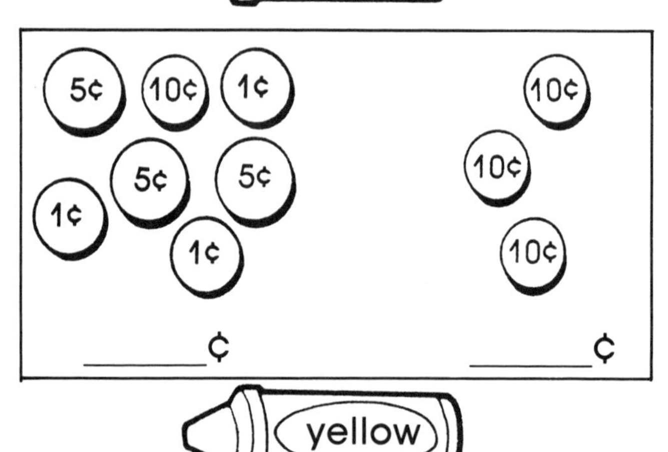

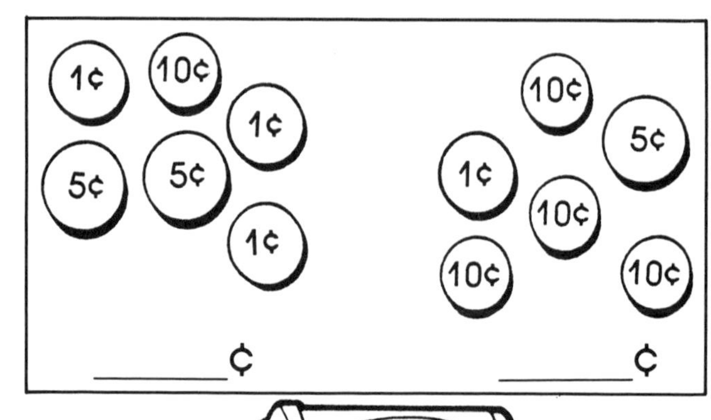

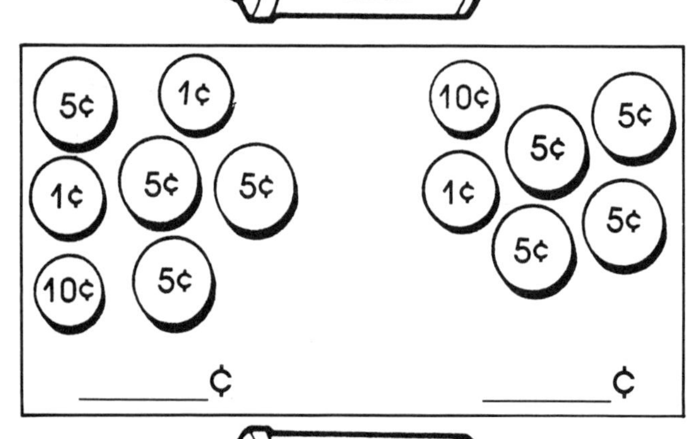

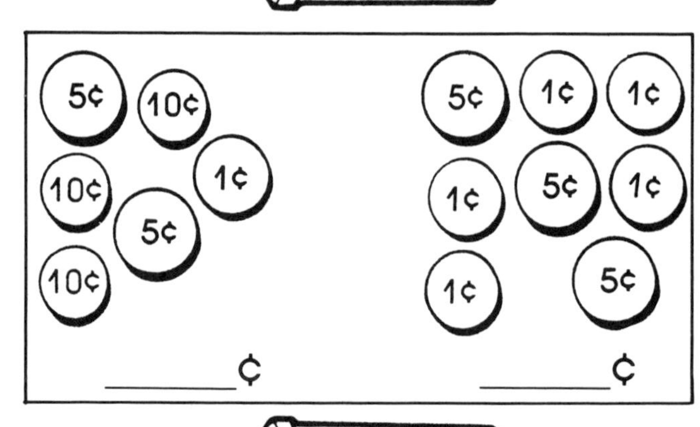

Smart Shoppers

Skill: Counting exact change

Name ______________________________

Circle the coins to equal the right amount.

Example

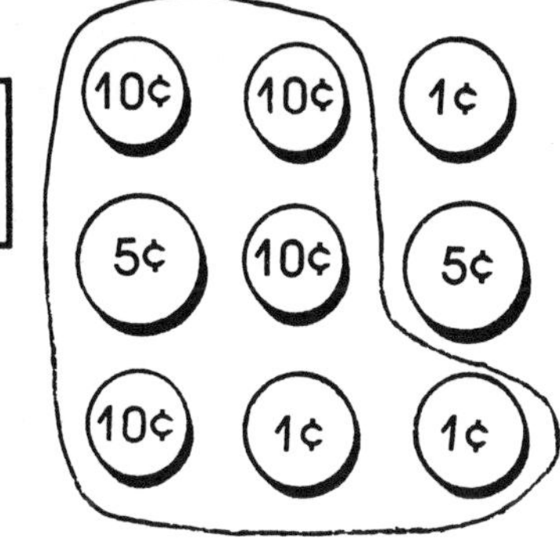

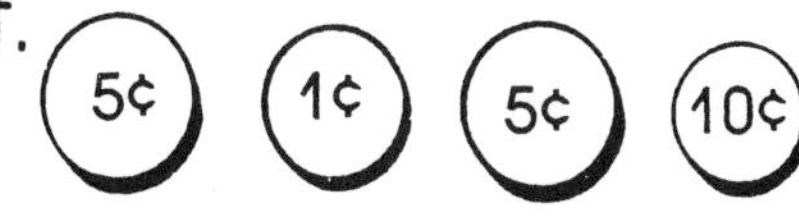

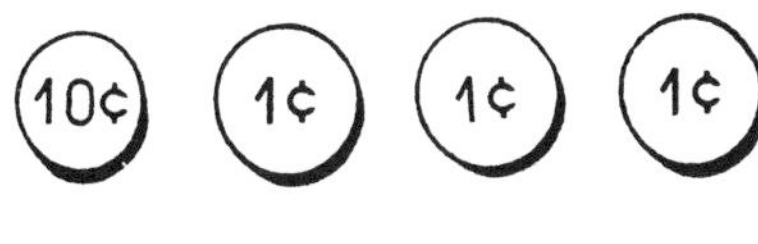

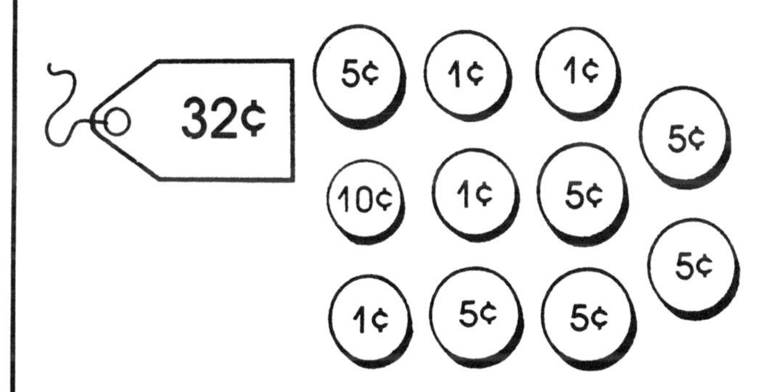

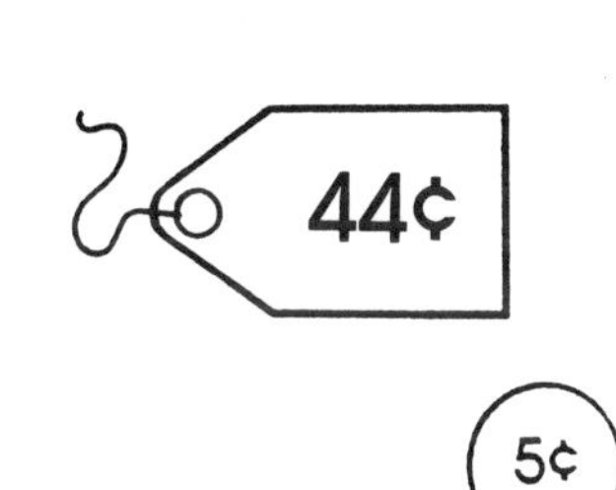

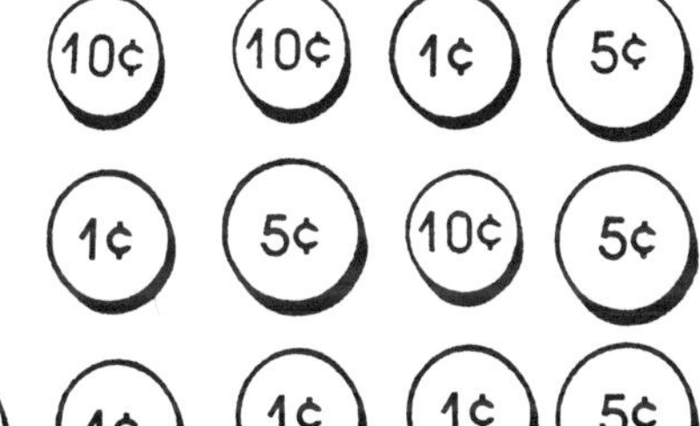

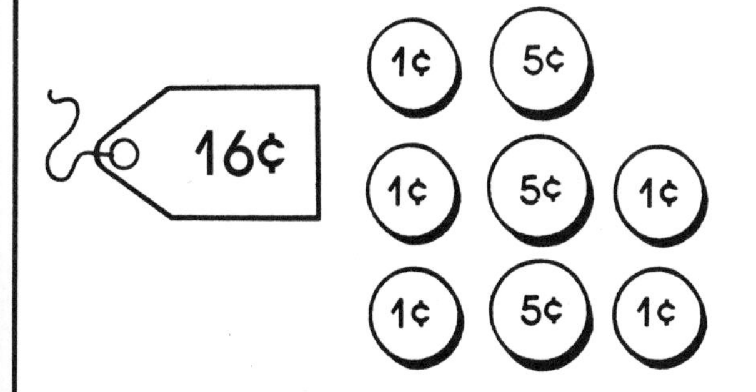

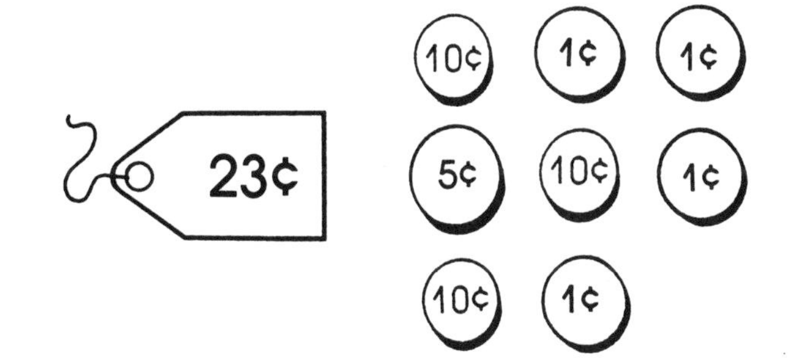

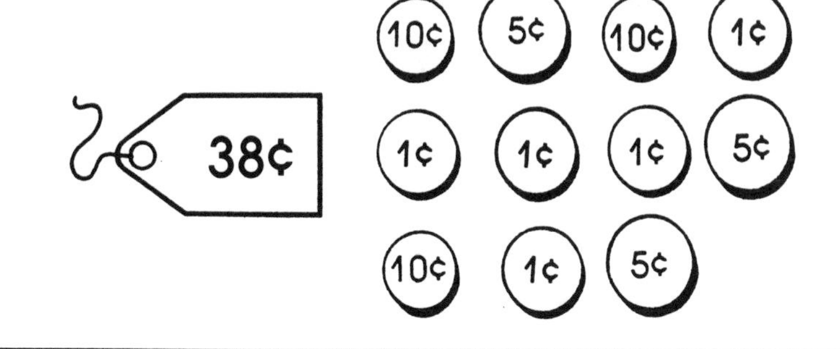

Join the Coins

Name ______________________________

Draw a line from the coins to the right amount.

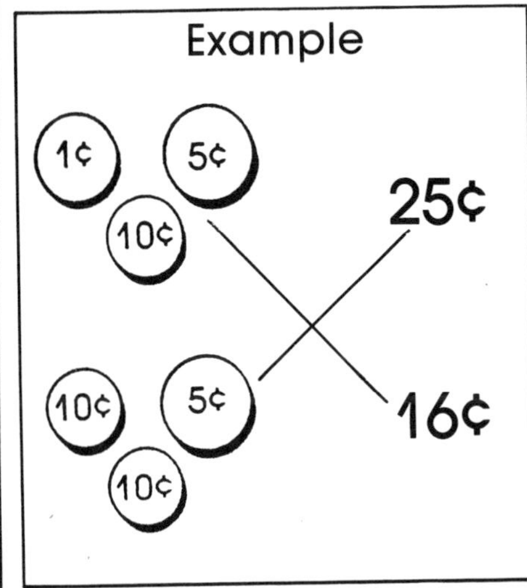

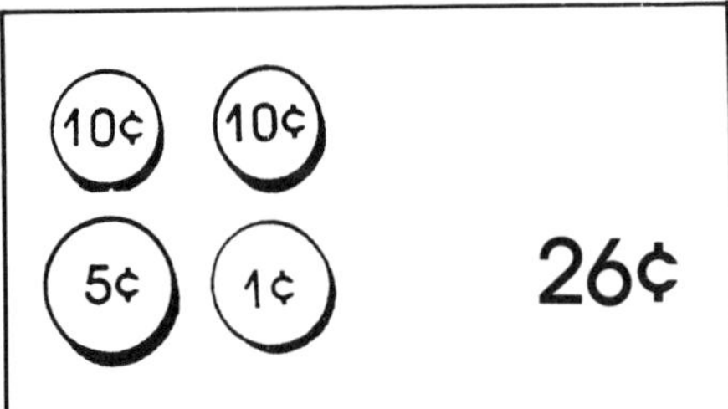

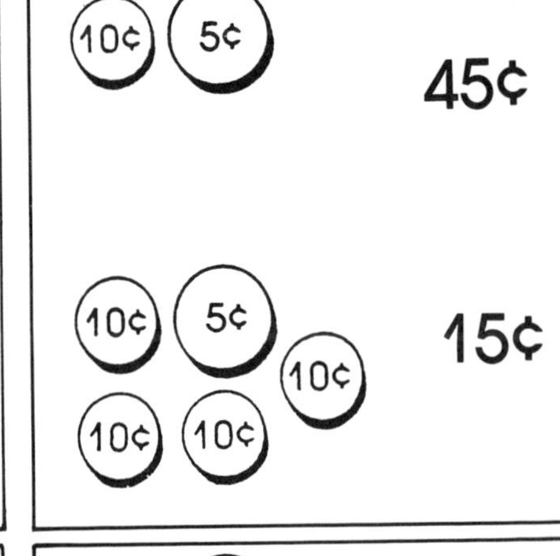

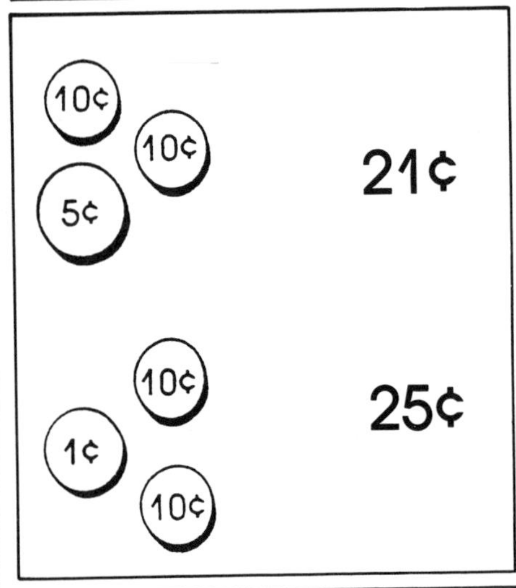

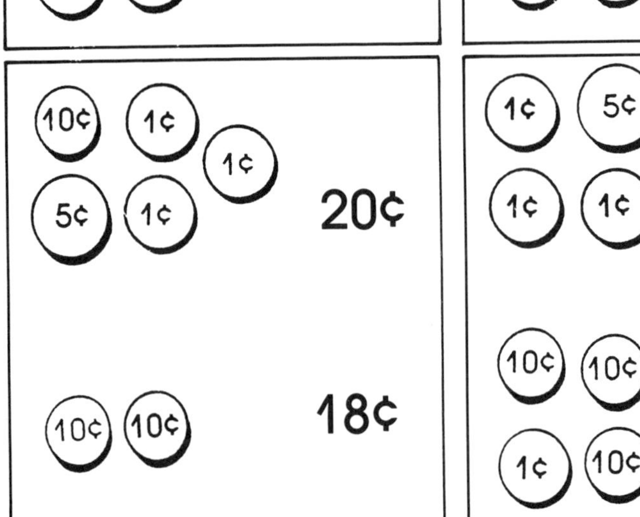

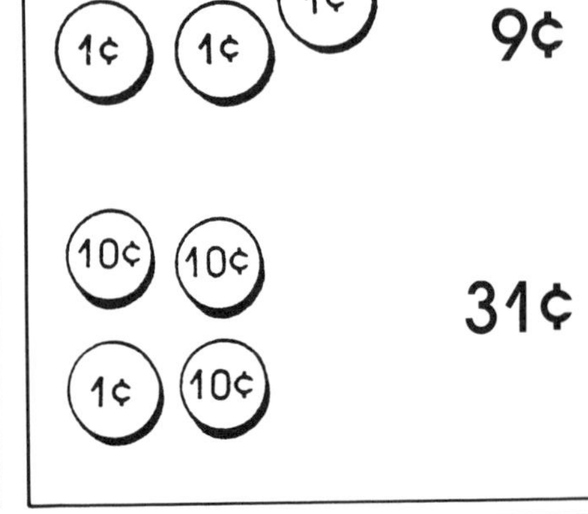

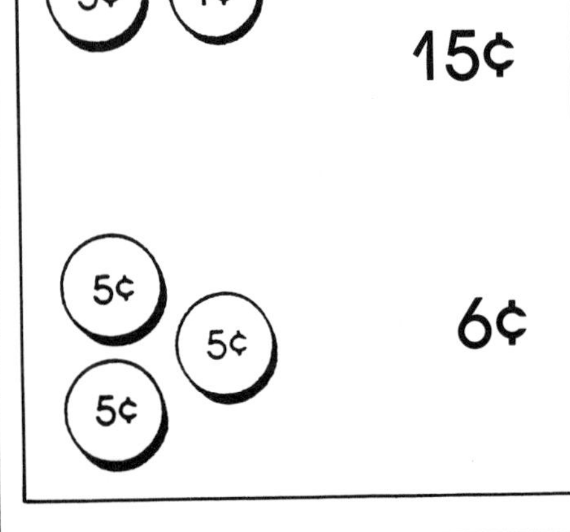

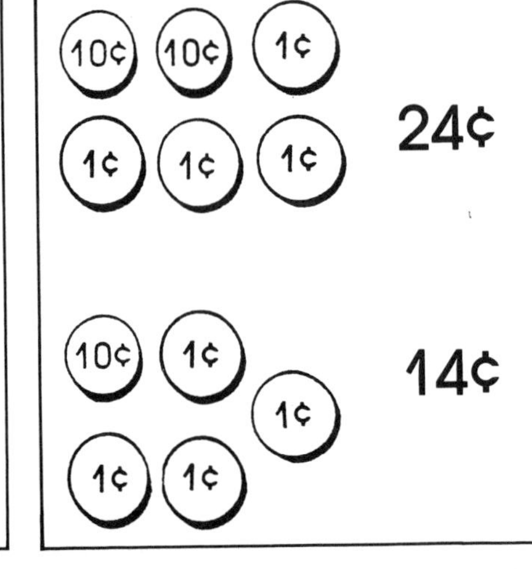

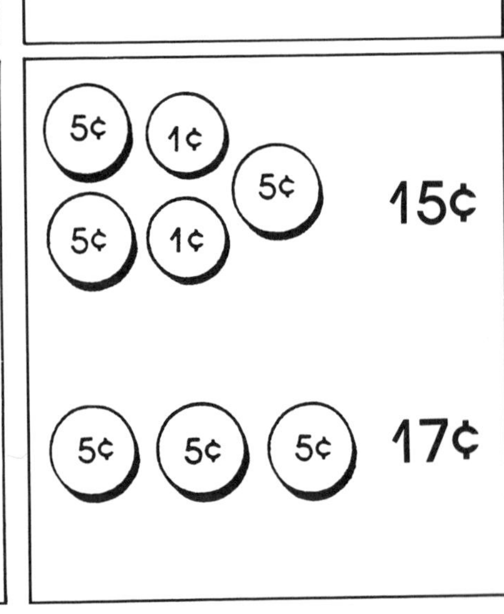

Skill: Counting money – dime, quarter

Quarterback Attack

Name ______________________

Count the coins. Write the amount in the football.

Example

25¢ 10¢ 10¢ 45¢

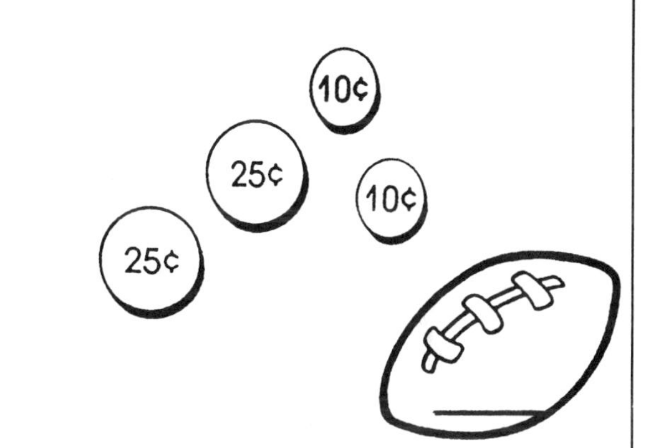

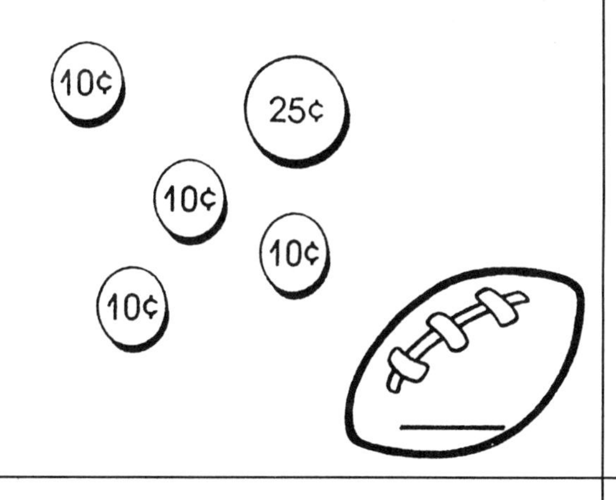

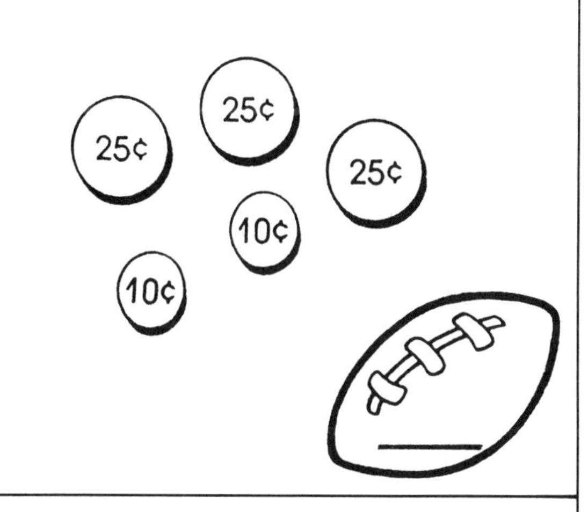

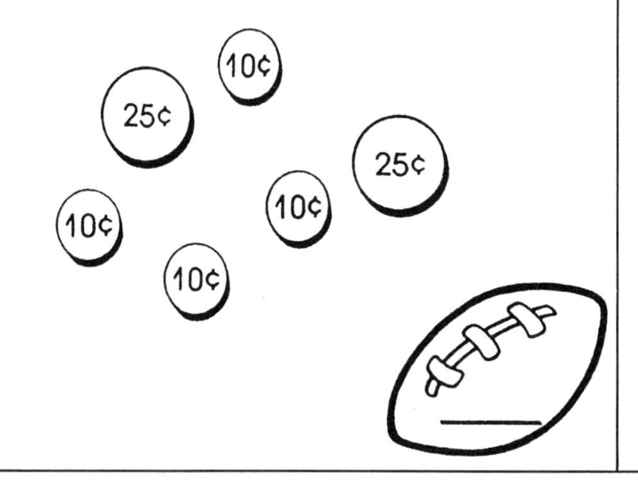

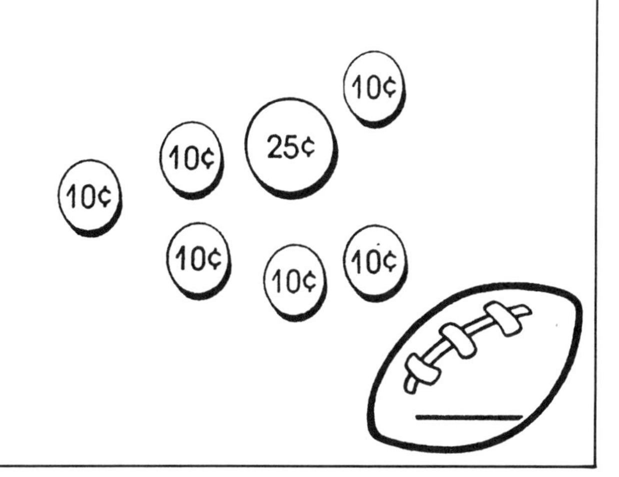

Skill: Money – comparing value

You Make "Cents"

Name ____________________

Count the coins.
Do you have enough money to buy each toy?

Example

	Cost	You have...		yes or no
	 58¢	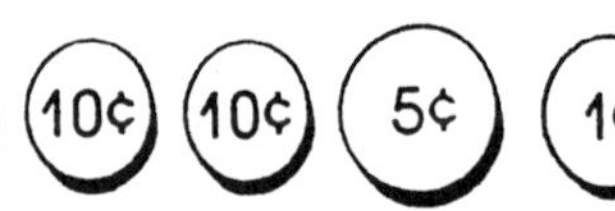 25¢ 10¢ 10¢ 5¢ 1¢	= 51¢	no
	 47¢	  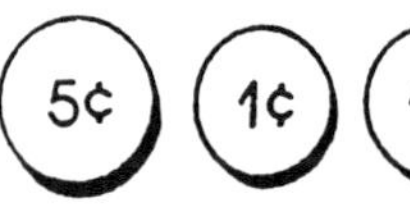 25¢ 25¢ 5¢ 1¢ 1¢	= ______	______
	75¢	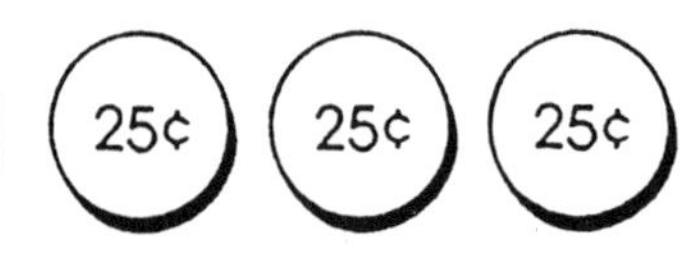 25¢ 25¢ 25¢ 1¢	= ______	______
	 43¢	 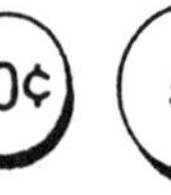 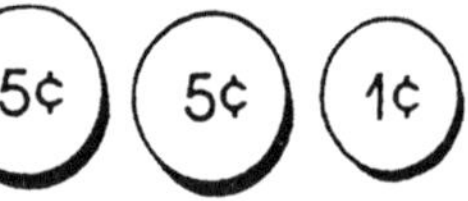 10¢ 10¢ 10¢ 5¢ 5¢ 1¢	= ______	______
	98¢	 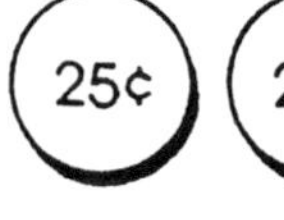  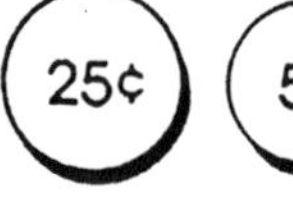 25¢ 25¢ 25¢ 25¢ 5¢	= ______	______
	 32¢	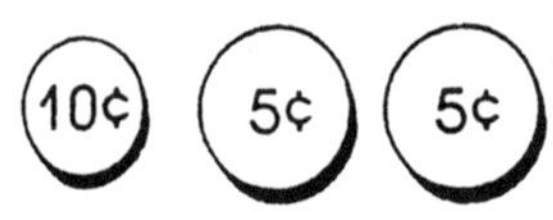 10¢ 5¢ 5¢ 5¢	= ______	______
	26¢	 25¢ 1¢ 1¢ 1¢ 1¢	= ______	______
	45¢	 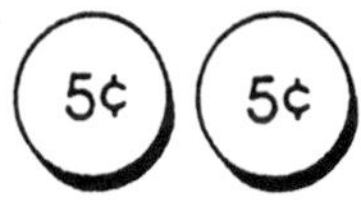 25¢ 5¢ 5¢	= ______	______

Skill: Money – using the decimal points and $. ($.01 – $2.00)

A Good Match

Name ______________________________

Count the money.
Draw a line to match.

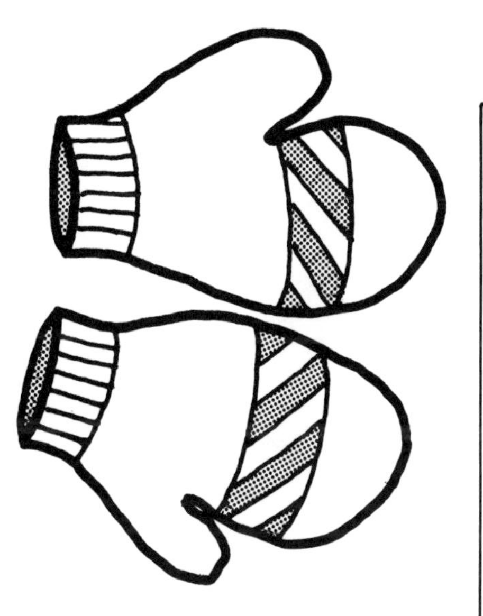

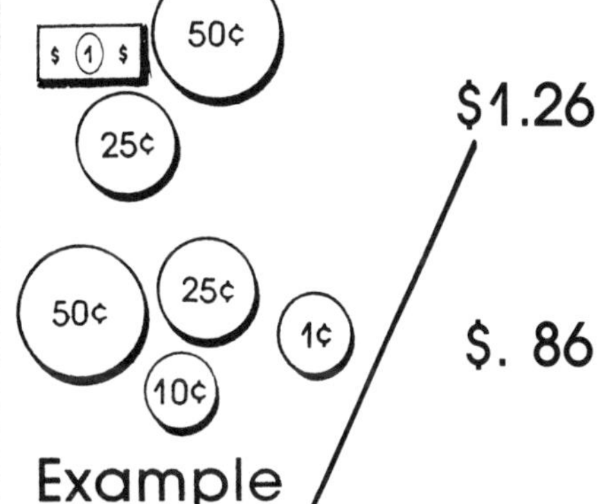

$1.26

$. 86

$1.75

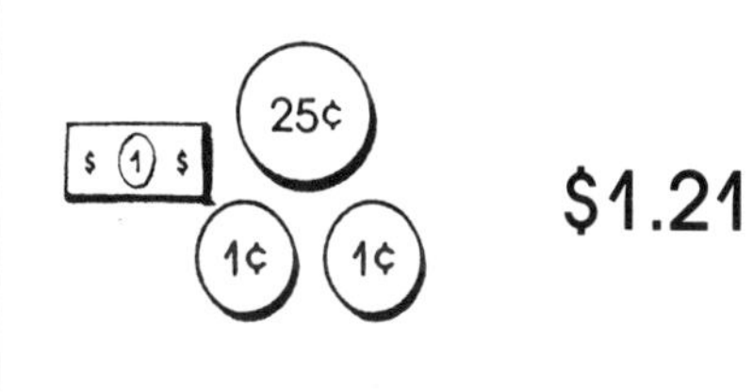

$1.27

$1.21

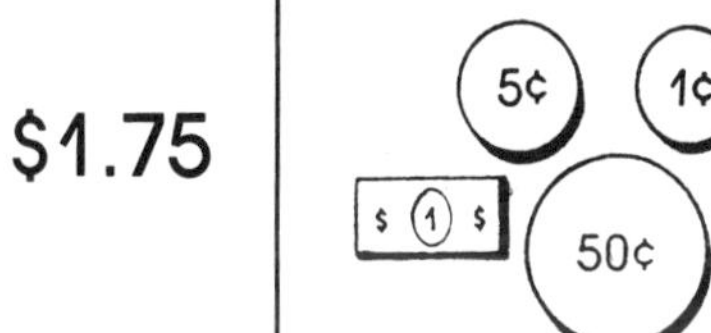

$1.56

$1.03

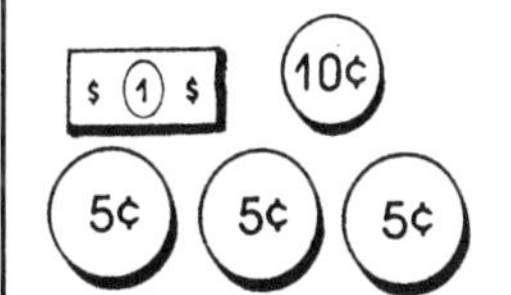

$1.25

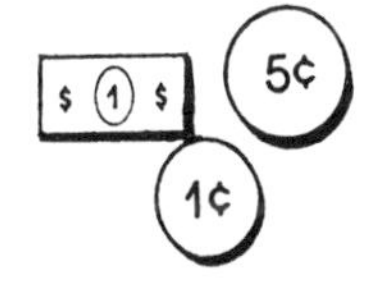

$1.65

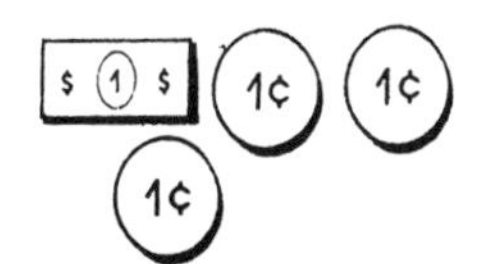

$1.10

$1.02

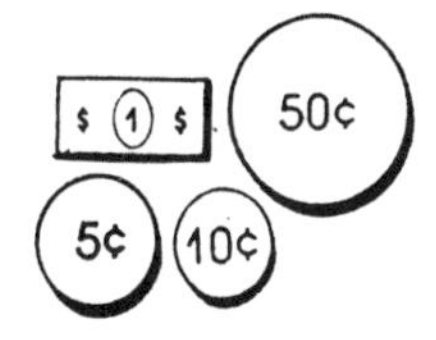

$1.06

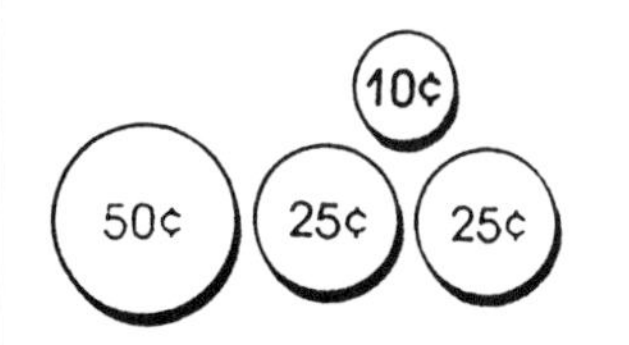

$1.07

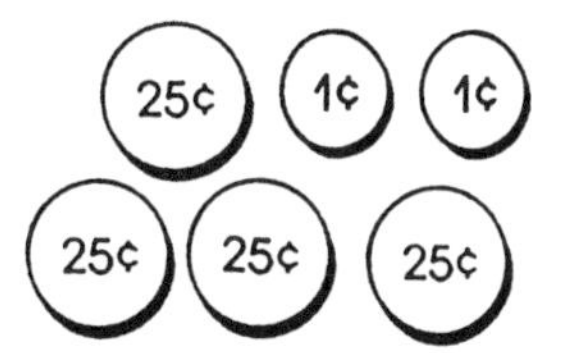

$1.01

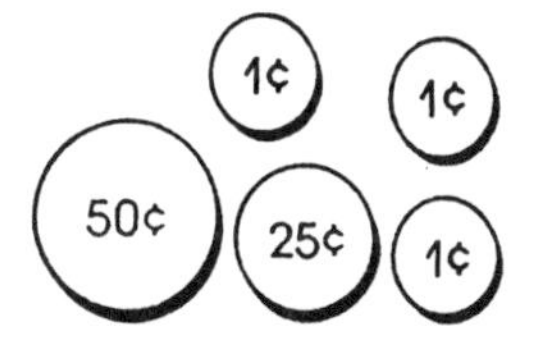

$ .78

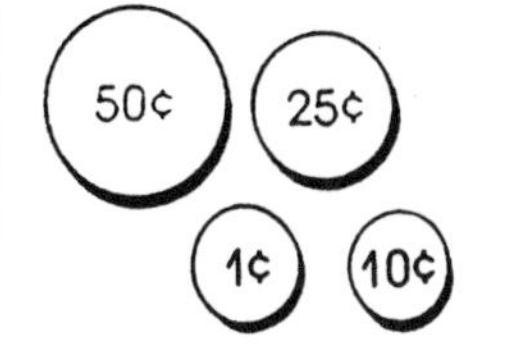

$1.01

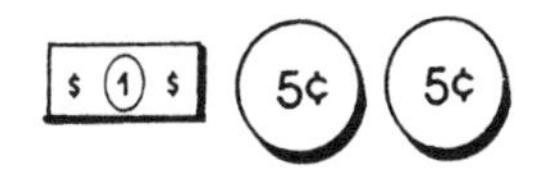

$1.10

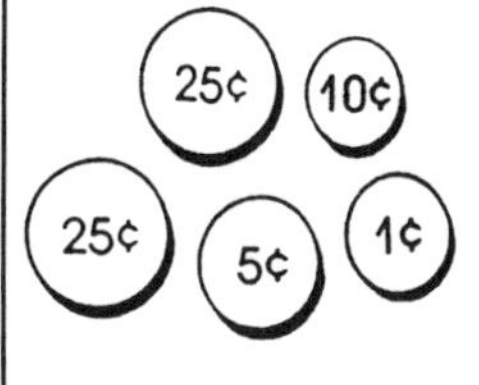

$ .75

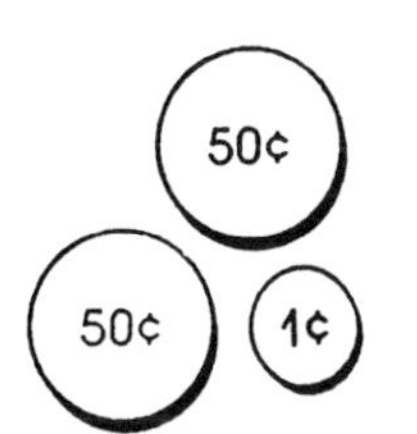

$ .86

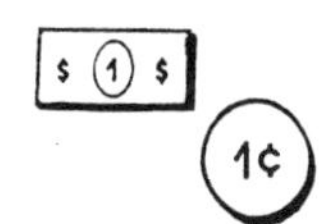

$1.01

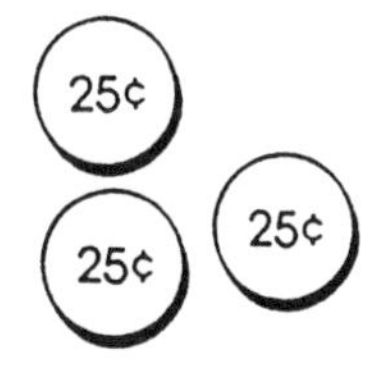

$ .66

Skill: Making change
Writing money 2 ways

Changes

Name ______________________

How much change should you get?

Example

Bought: 29¢ + 14¢ = 43¢

I have: 50¢, 5¢

55¢
− 43¢
12¢

Change 12¢

Bought: 56¢ + 27¢

I have: $1, 25¢

Change

Bought: 61¢ + 59¢

I have: 25¢, $1, $1

Change

Bought: $ .78 + $ .69

I have: $1, $1, 50¢, 25¢

Change

Bought: $ .59 + $ .86

I have: $1, 25¢, 50¢

Change

Bought: $ .66 + $ .75

I have: $1, $1, $1

Change

Lunch Money

Skill: Making change (to $3.50)
Writing money 2 ways

Name ______________________

How much change?

Example

Lunch

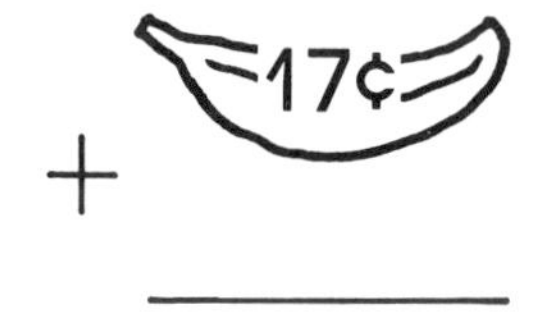

\+ ______

76¢

I have

$1 10¢ 5¢

115¢
− 76¢
______
39¢

Change

Lunch

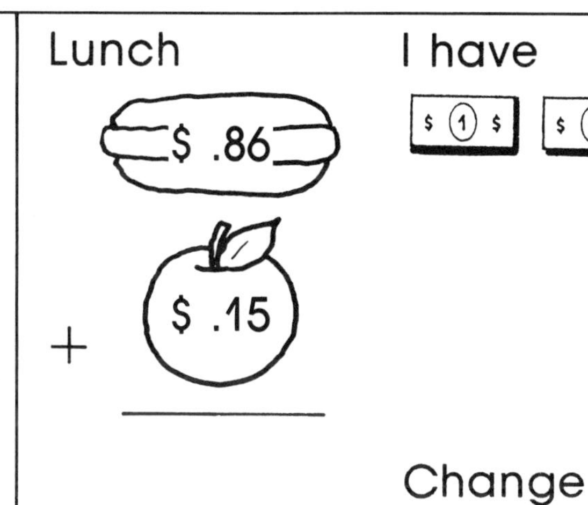

\+ ______

I have

Change [ ]

Lunch

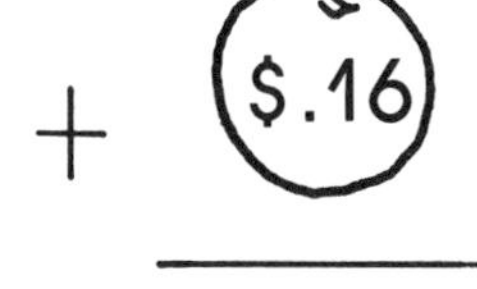

\+ ______

I have

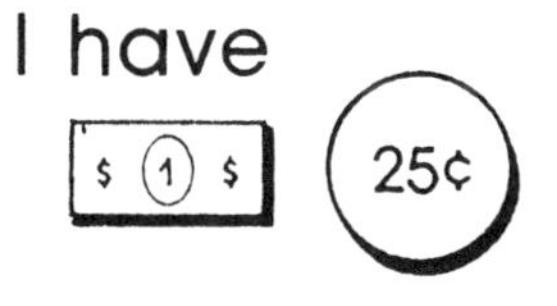

Change

Lunch

\+ ______

I have

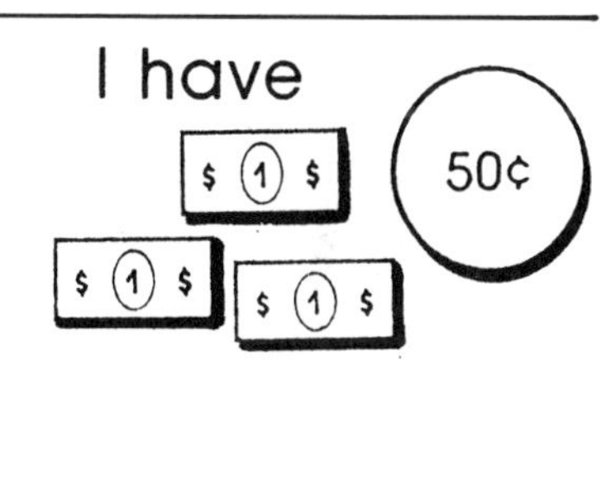

Change

Lunch

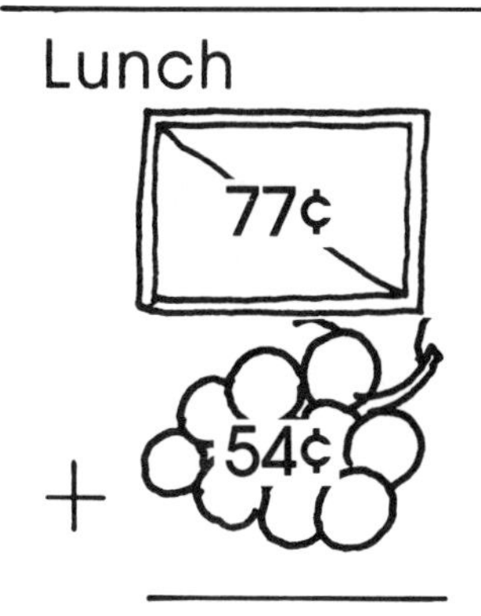

\+ ______

I have

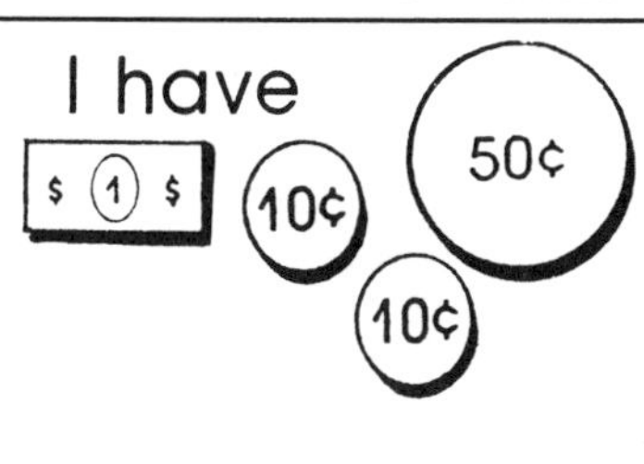

Change [ ]

Lunch

\+ ______

I have

$1 $1 $1

Change [ ]

Go Bananas Over Money!

Name ______________________________

Write an addition sentence for each problem.

Example

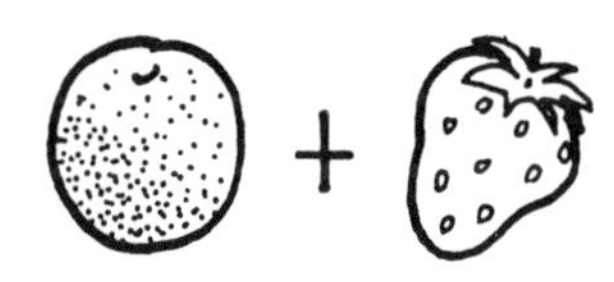

7¢ + 5¢ = 12¢

Skill: Telling time – hour

Hickory Dickory Dock

Name ____________________

Skill: Telling time – hour

It's Clown Time!

Name ____________________

What time is it?

Here's the Scoop!

Skill: Telling time – hour

Name ______________________________

Draw the hour hand on each clock.

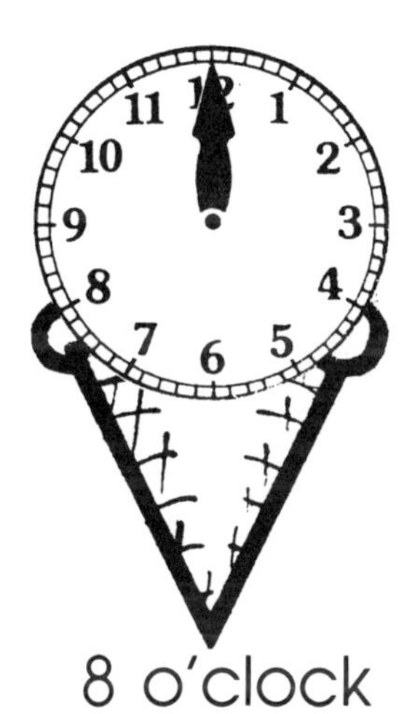

8 o'clock

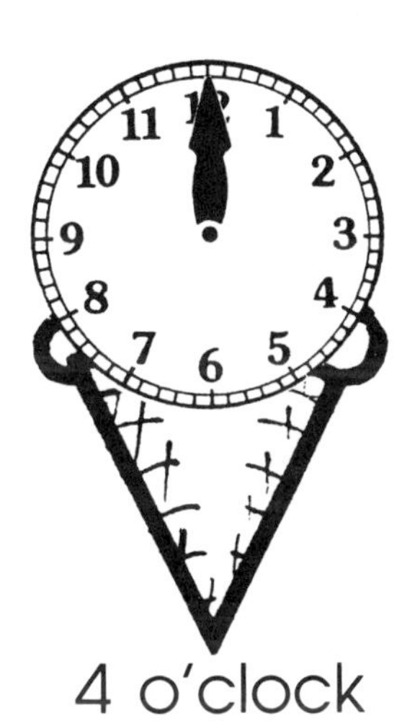

4 o'clock

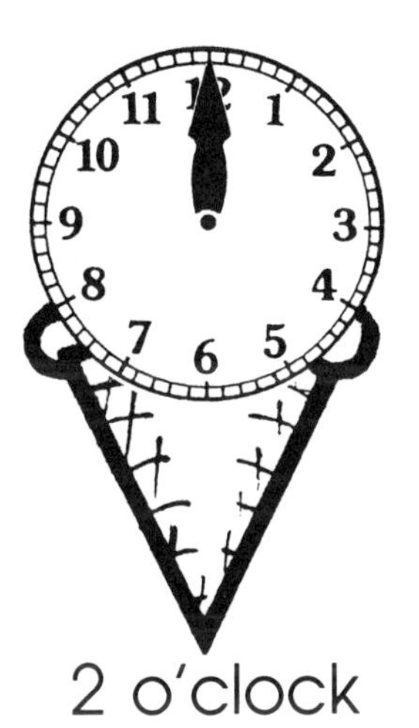

2 o'clock

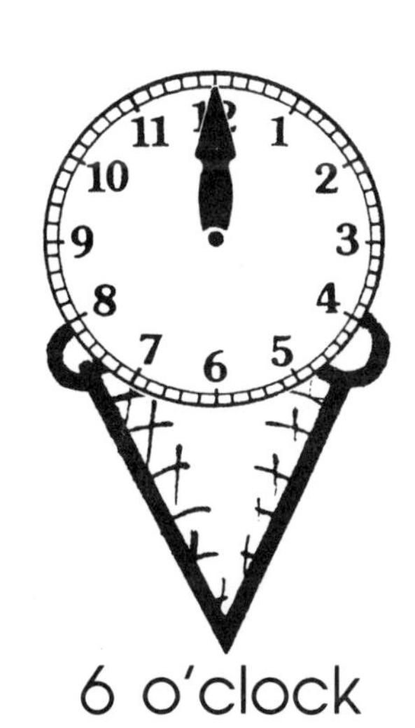

6 o'clock

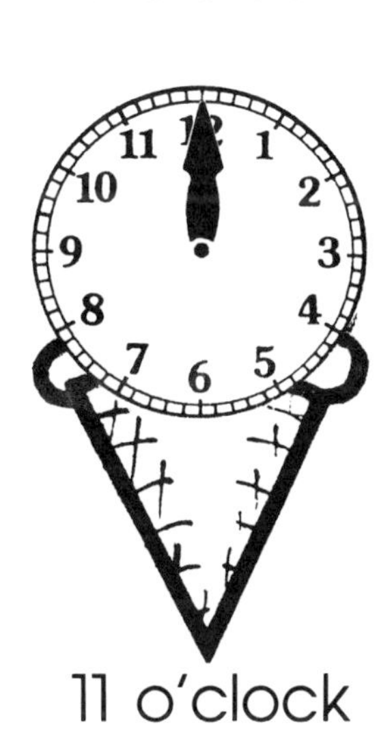

11 o'clock

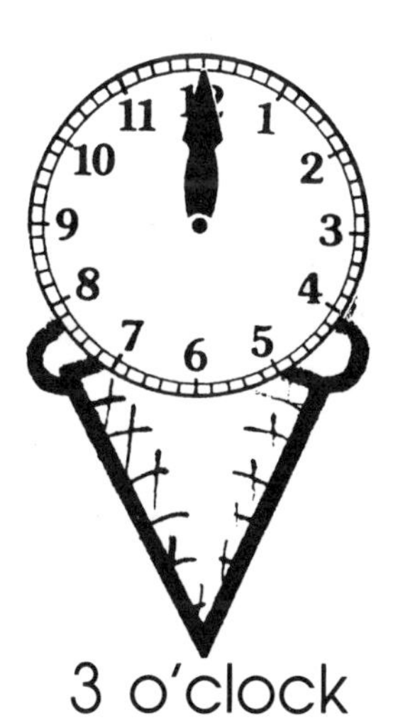

3 o'clock

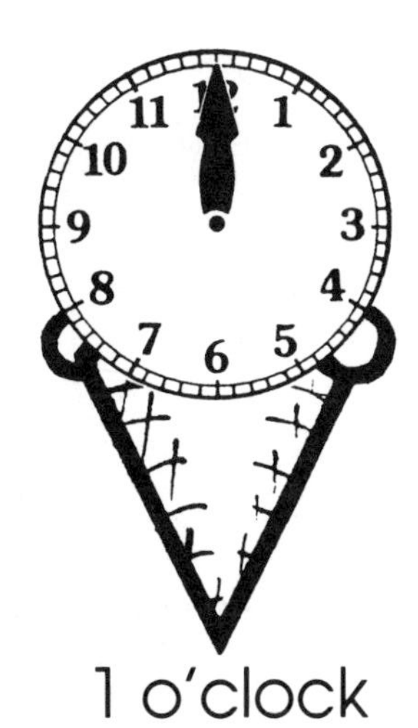

1 o'clock

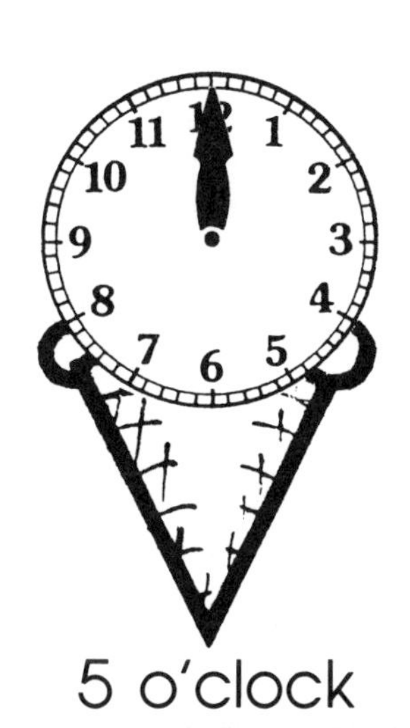

5 o'clock

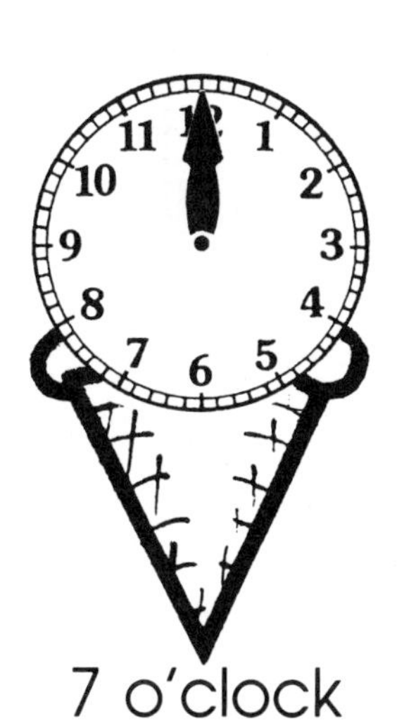

7 o'clock

Skill: Telling time – hour, half-hour

Who "Nose" These Times?

Name ____________________

Write the time under each clock.
Color the noses.

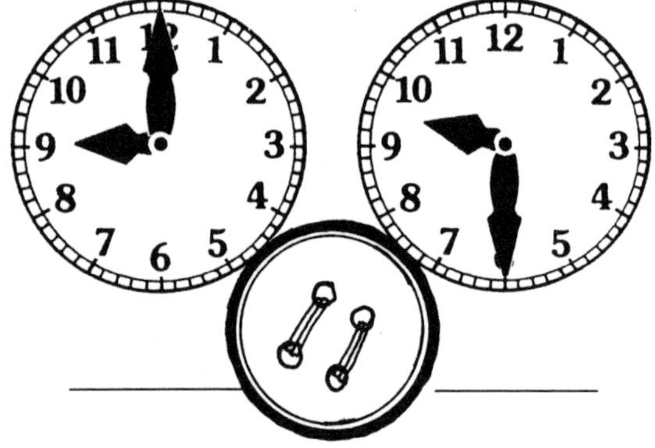

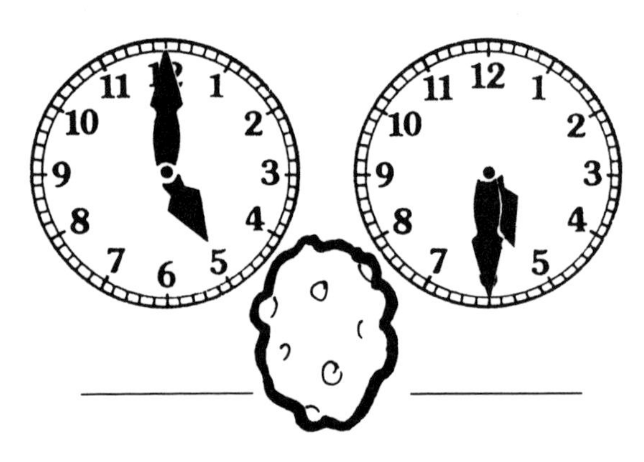

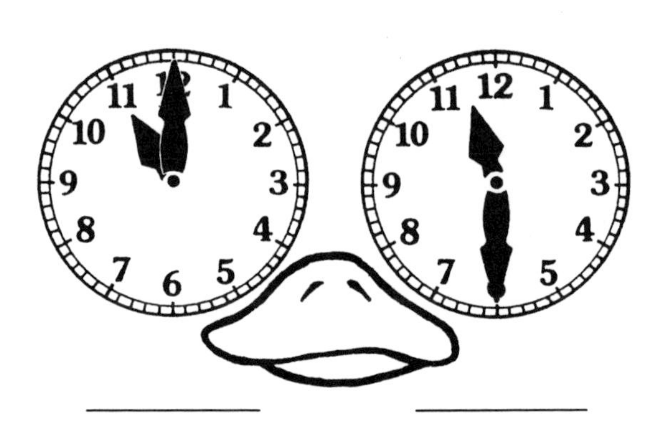

Skill: Telling time – hour, half-hour

Space Time

Name ____________________

What time is it?

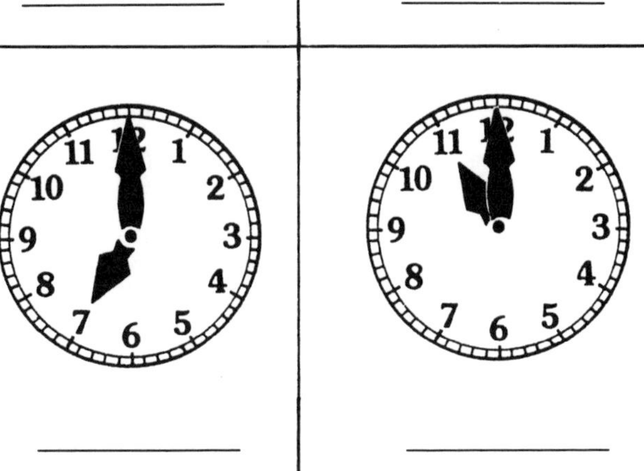

Skill: Telling time – hour, half-hour

You Are My Sunshine

Name ______________________

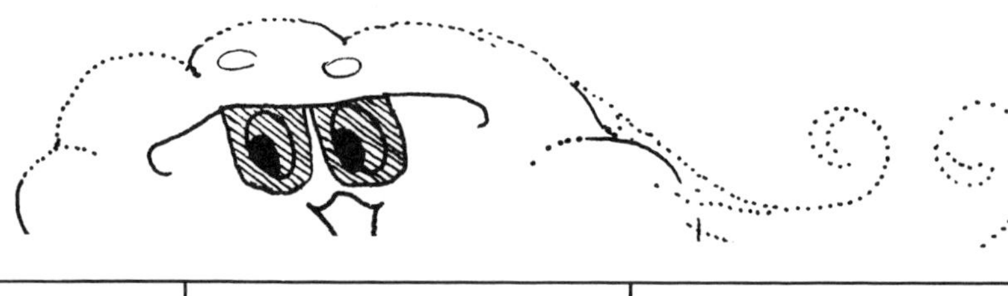

What time is it?

	 ________	 ________	 ________
 ________	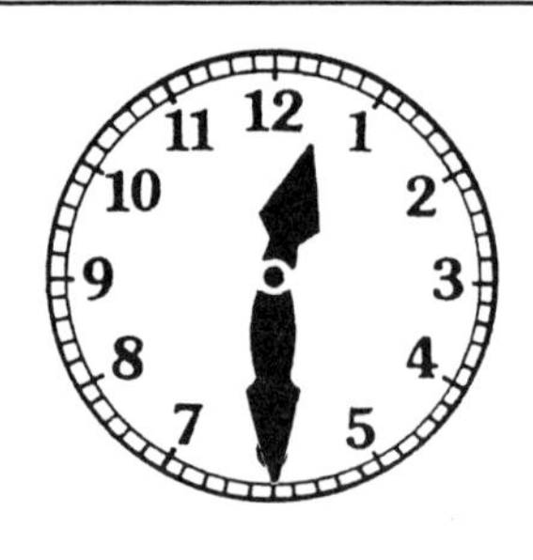 ________	 ________	 ________
 ________	 ________	 ________	 ________
 ________	 ________	 ________	 ________

Sock Clocks

Name ____________________

Draw the hands on the sock clocks.

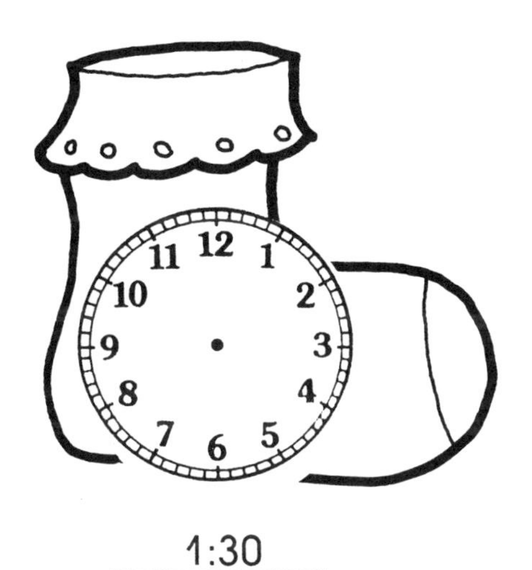

1:30

7:00

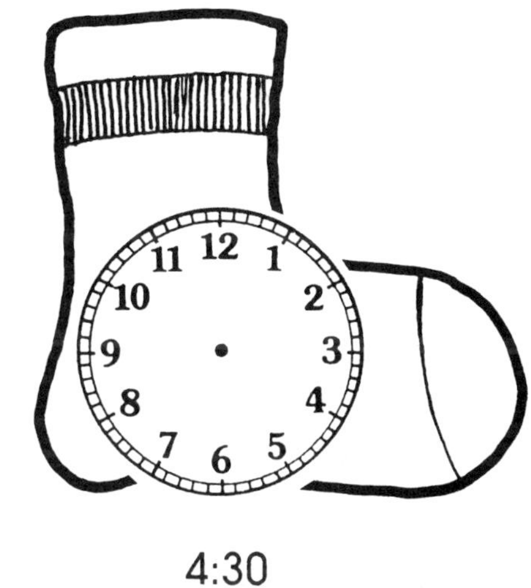

4:30

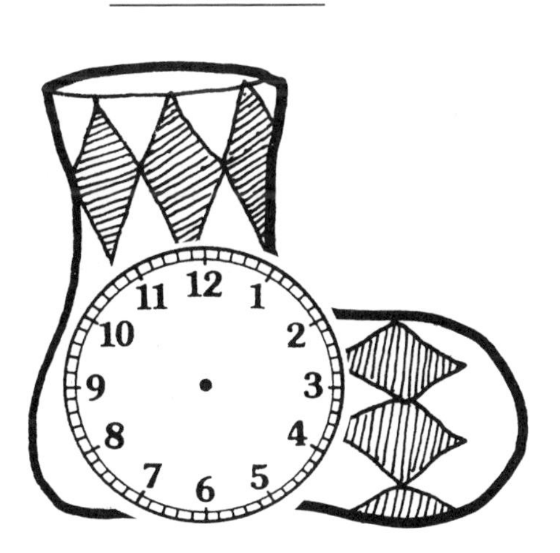

10:00

3:30

9:30

4:00

2:30

6:00

Time Lines

Name ____________________________

Match each clock to the correct time. Draw a line with the color by the clock.

red

brown

green

yellow

7:00

2:30

2:00

8:30

blue

red

9:30

1:30

orange

green

10:00

12:00

4:30

11:00

purple

blue

5:00

3:30

black

purple

My Family Time Tree

Name ________________________________

Write the time.
Draw the hands on each clock.

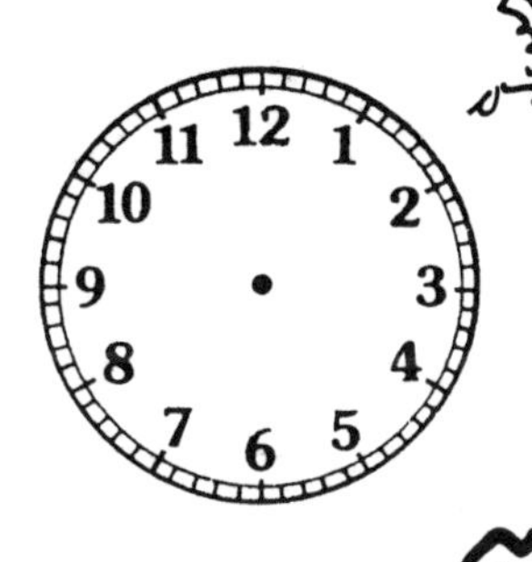

I get up at______.

I go to bed at______.

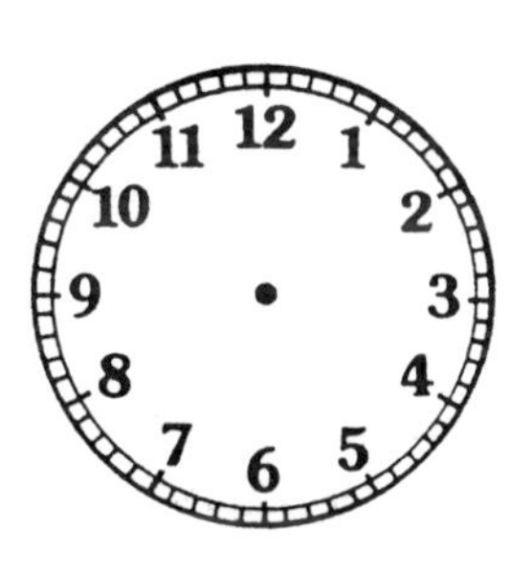

School starts at______.

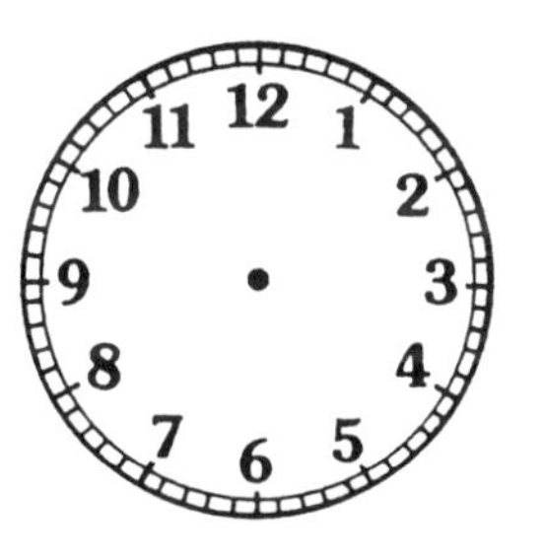

I watch TV at______.

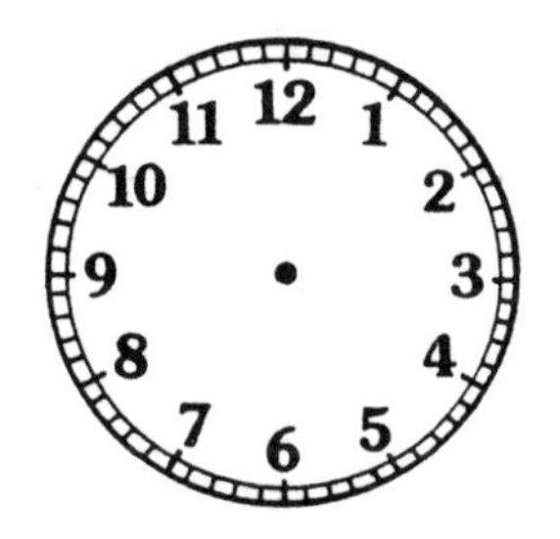

Lunch is at______.

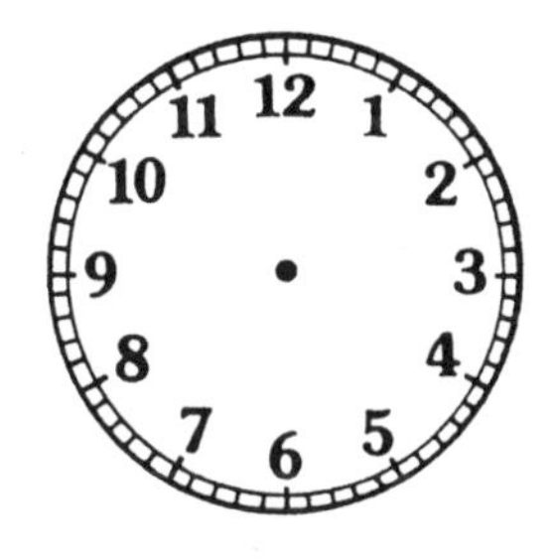

Dinner is at______.

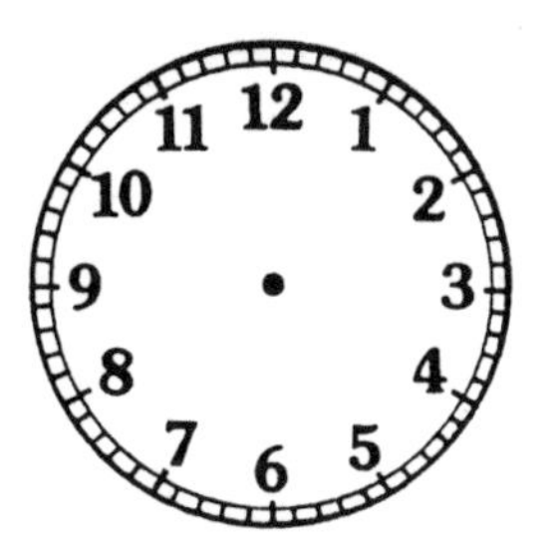

Recess is at______.

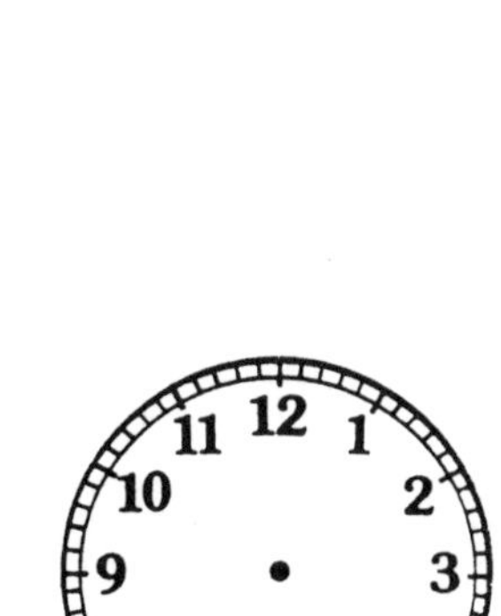

School ends at______.

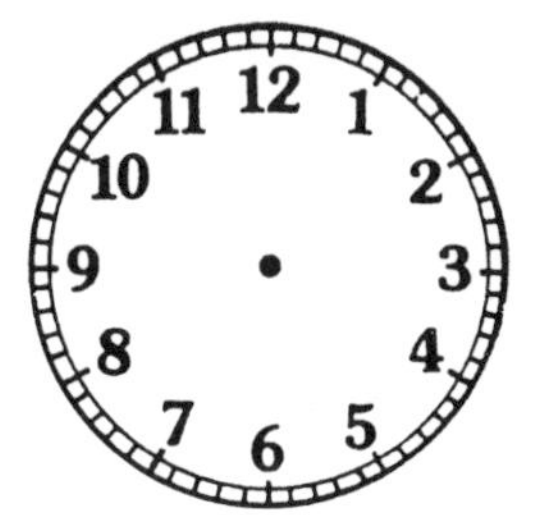

I play at______.

Turtle Time

Name ______________________________

What time is it?

Skill: Telling time – 5 minutes

Fish Time

Name ____________________

Draw the hands on these fish clocks.

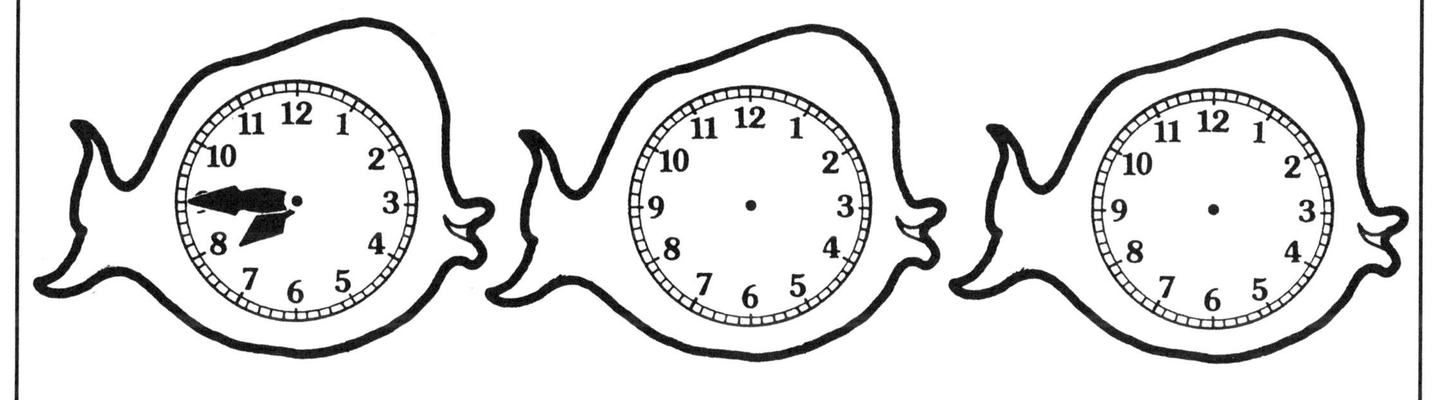

7:45 8:05 11:15

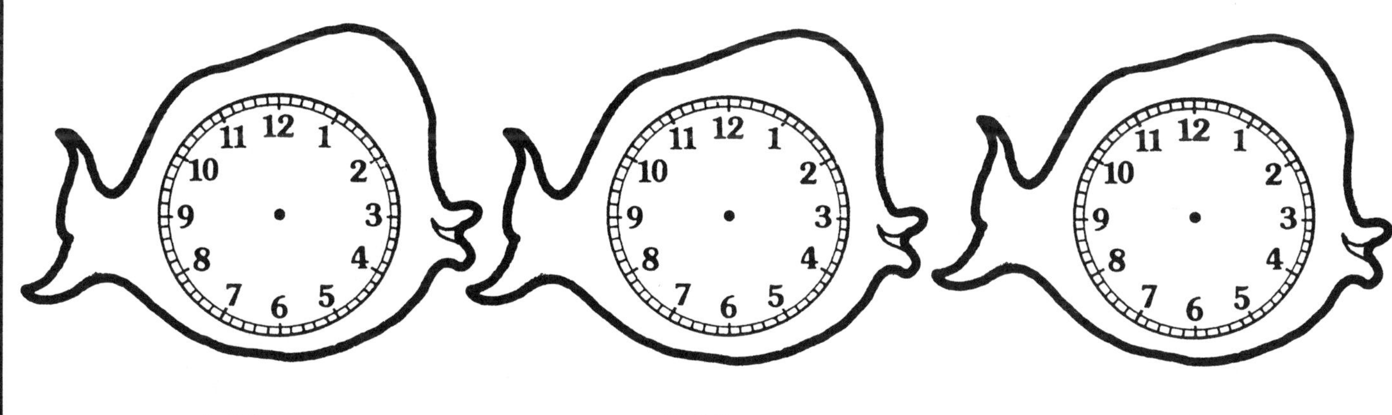

3:20 5:55 1:50

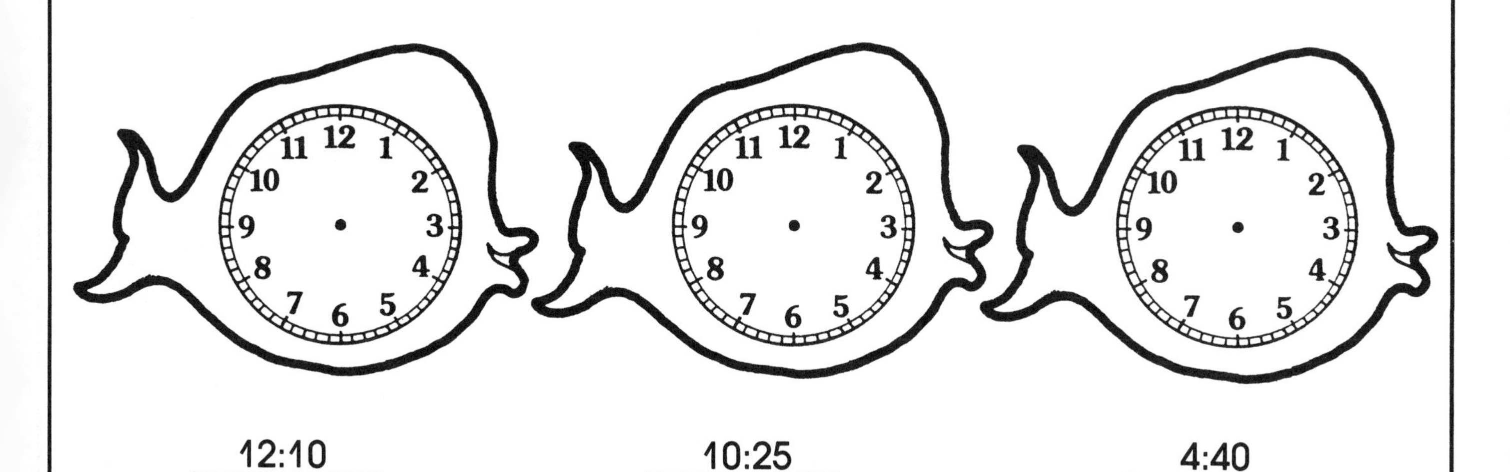

12:10 10:25 4:40

Clock Tower Times

Name ____________________

Cut out and paste under the correct clock.

12:35	8:40	4:15	9:55	10:05
6:10	1:45	12:20	2:45	3:25

Cartoon Time!

Name ______________________________

Draw the clock hands to show the time you watch these crazy cartoons.

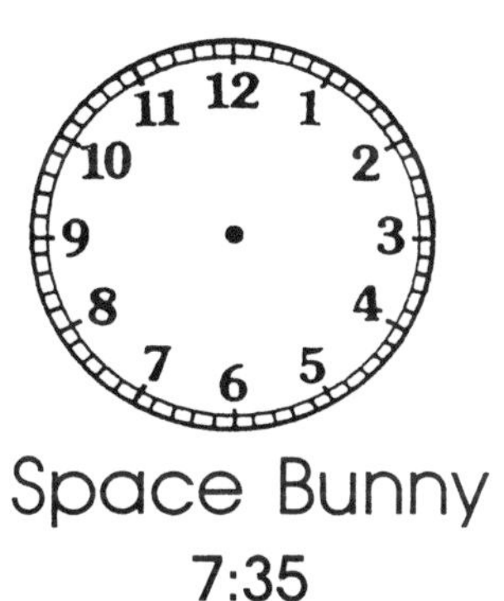

Space Bunny
7:35

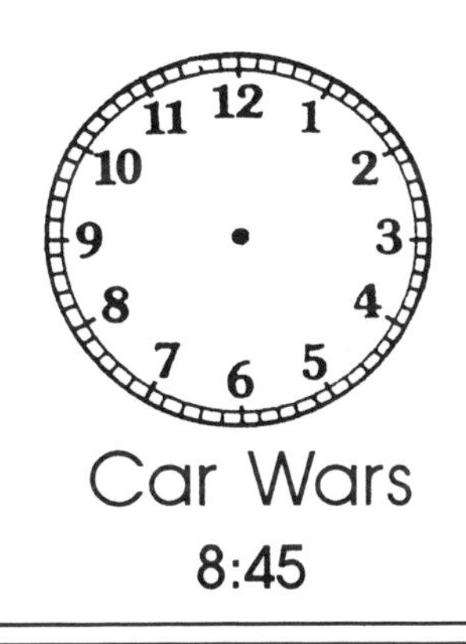

Car Wars
8:45

The Snuffs
5:15

Fun Runner
9:00

Scare Bears
2:40

Magic Elf
11:30

Tummy Bears
3:20

Monster Time
12:10

Sunny Funnies
1:05

What is your favorite cartoon? ______________________________

What time does it come on? ______________________________

Apple Time

Name ______________________________

Write the times on the worms.

Skill: Telling time – 5 minutes (digital)

As Easy as 1,2,3

Name ____________________

Draw the hands. Write the time.

Three thirty		3 : 30
Five forty-five	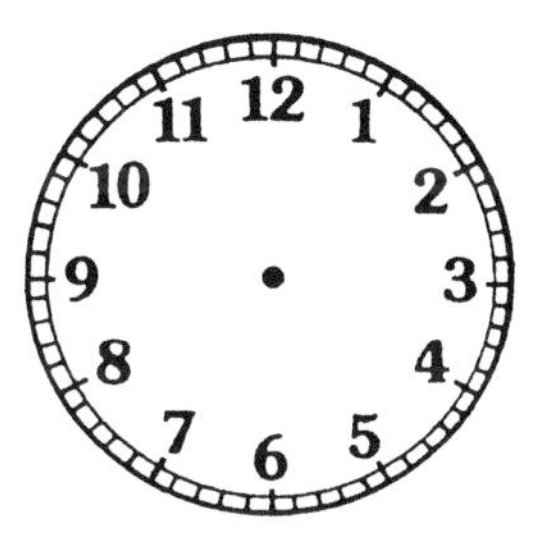	:
Eleven twenty	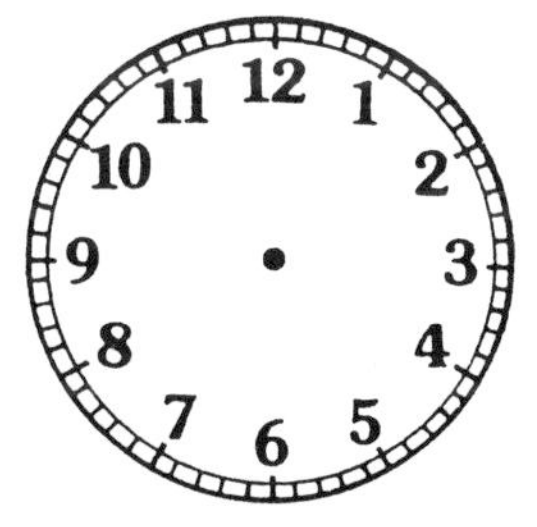	:
Eight ten	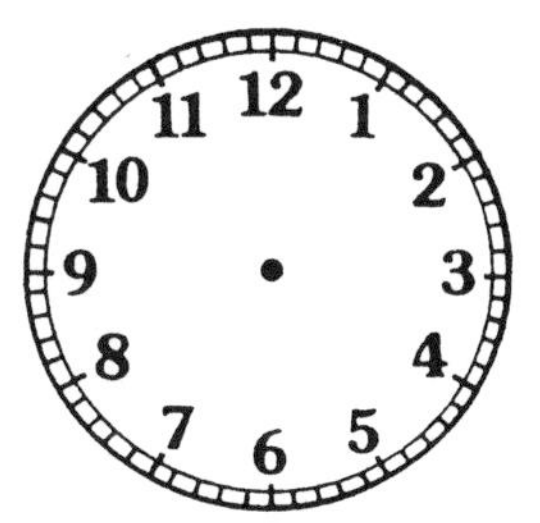	:
Two fifty-five		:
Nine forty	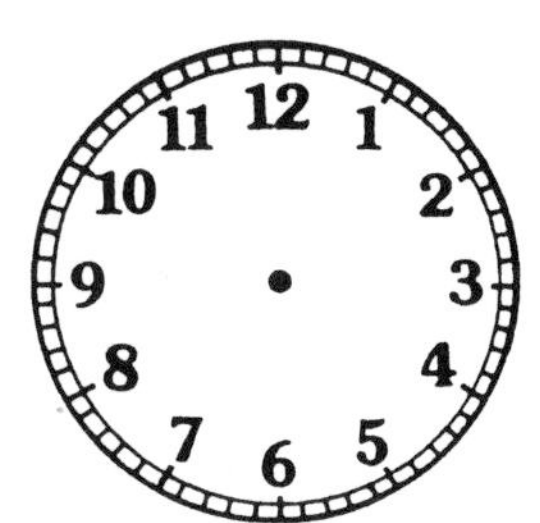	:

Skill: Telling time – 5-20 minute intervals.

"Watch" Me!

Name ______________________________

Add 10 minutes to each clock in this row.

8:30 ______ ______

Add 15 minutes to each clock in this row.

  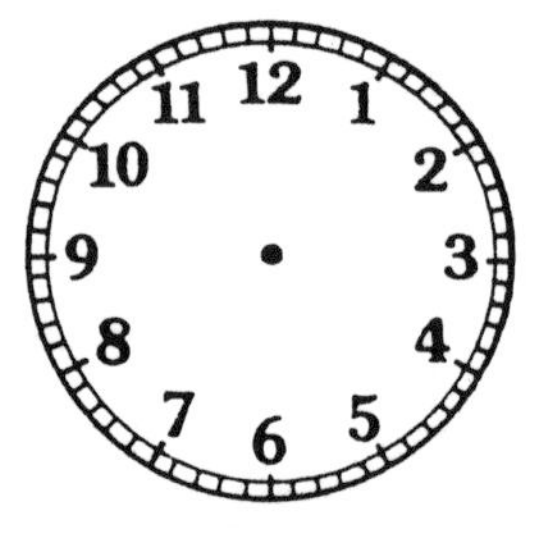

3:00 ______ ______

Add 5 minutes to each clock in this row.

 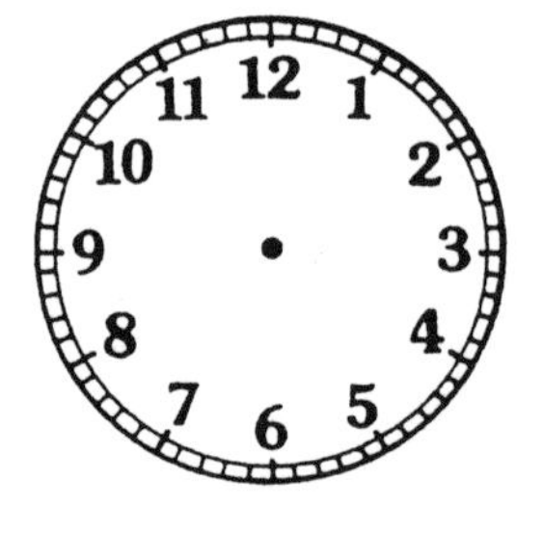

1:25 ______ ______

Add 20 minutes to each clock in this row.

 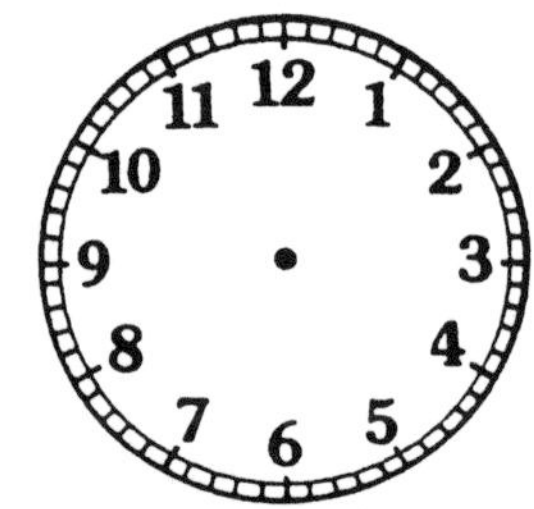 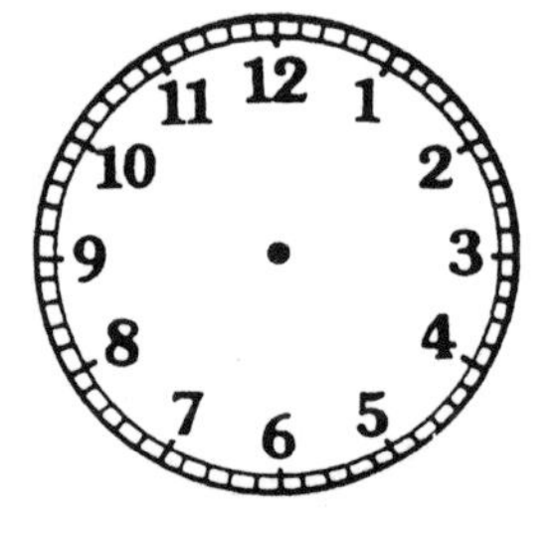

11:45 ______ ______

Turtle Spots

Name ____________________

Count the spots on the turtles.
Color the boxes to show how many spots.

Wormy Apples

Name ______________________________

Color the boxes to show how many worms.
Answer the questions.

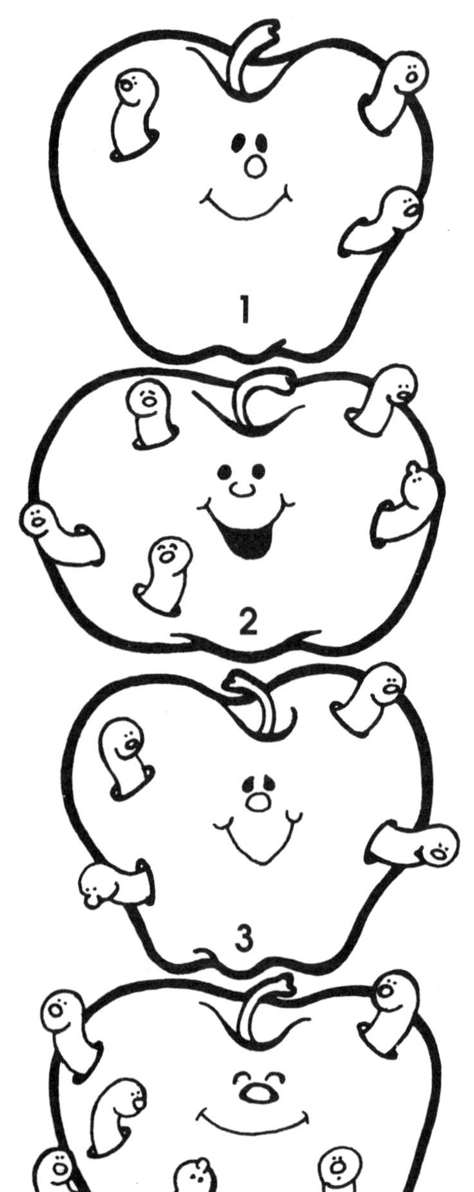

1	2	3	4	5	6

How many worms in apple 1?____ 2?____ 3?____ 4?____

In apples 1 and 3? ____ In apples 2 and 4? ____

How many more worms in apple 4 than in apple 2? ____

How many more worms in apple 3 than in apple 1? ____

Worm Graph

Name ____________________

Color the sums: 5-blue 6-red 7-yellow 8-orange 9-green

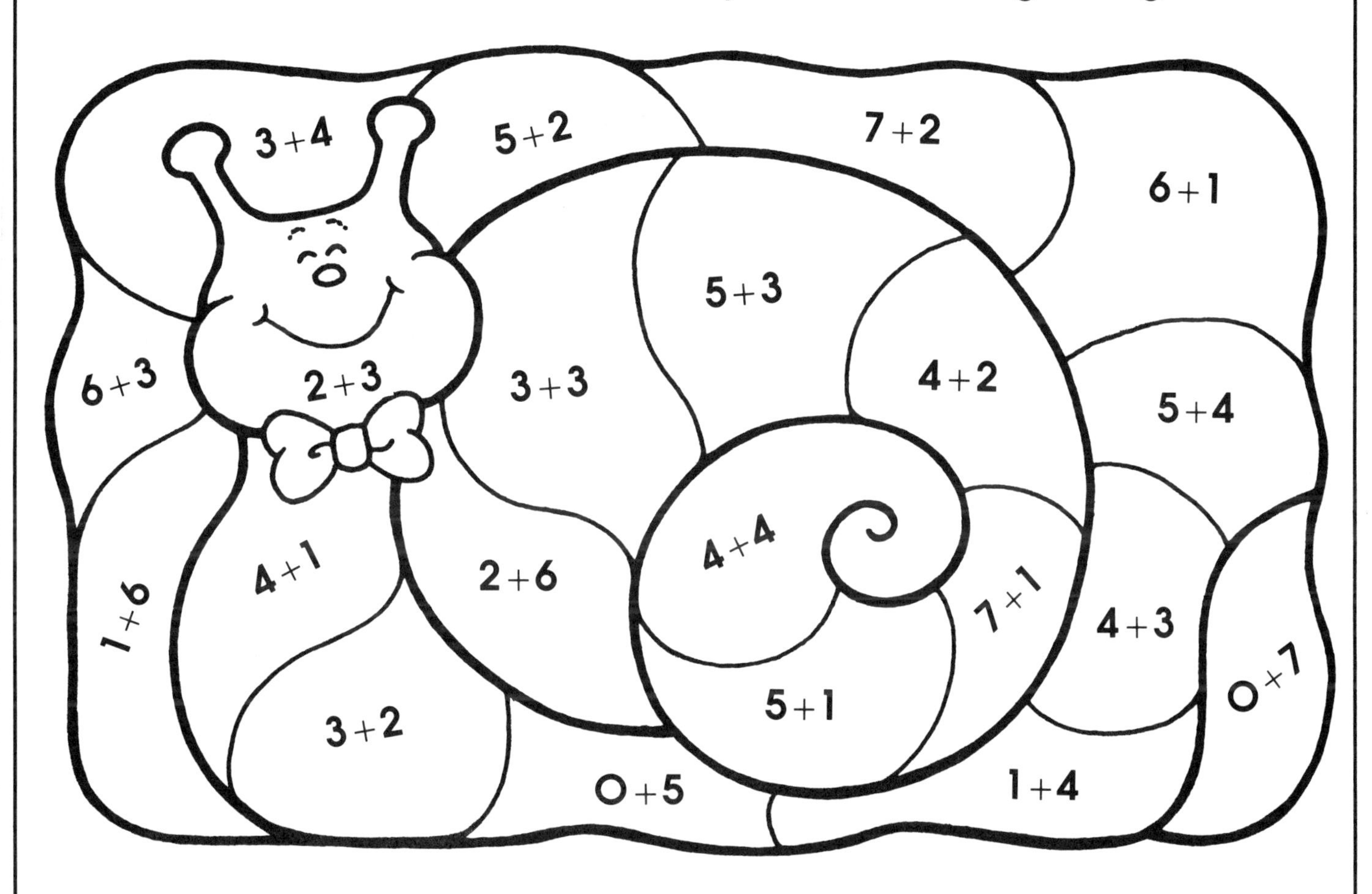

Color the squares on the graph to show how many.

	1	2	3	4	5	6
blues						
reds						
yellows						
oranges						
greens						

Color Graph

Name ____________________

How tall is each crayon?
Cut out the ruler and measure.

cm = centimeter

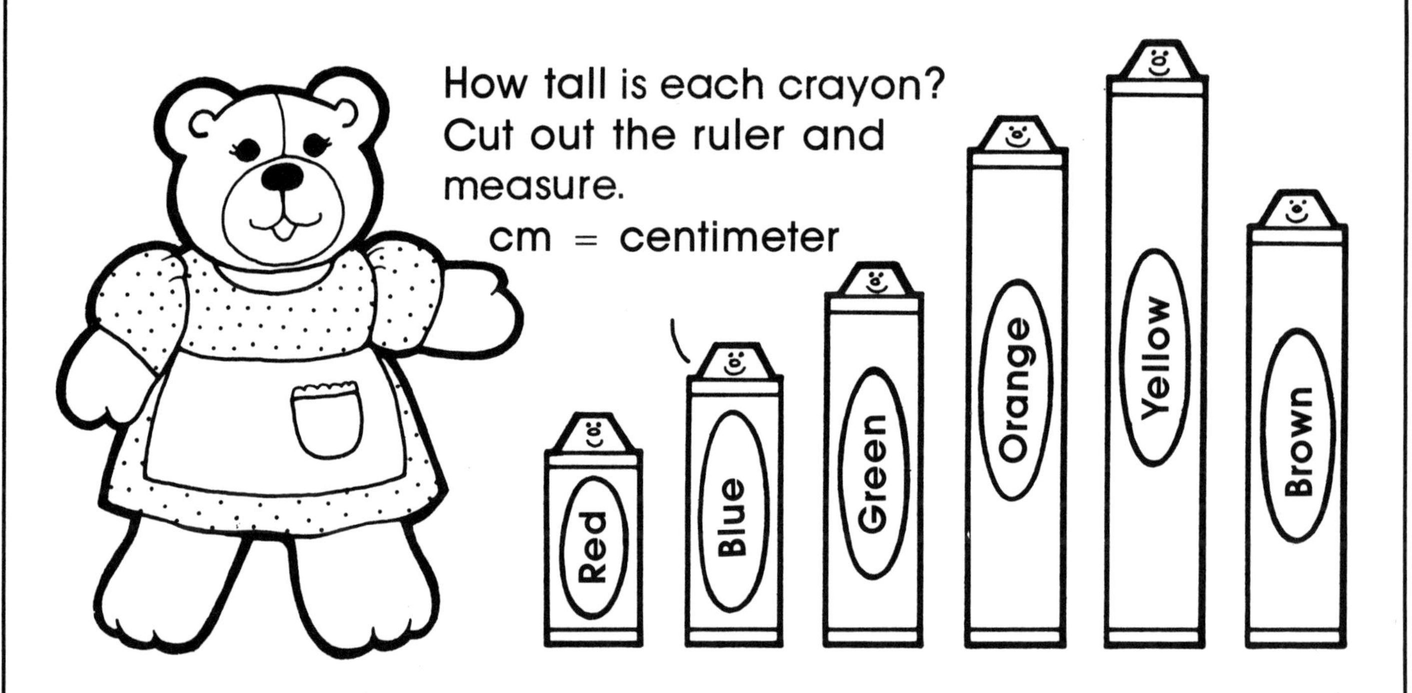

Color the squares on the graph to show how tall each crayon is.

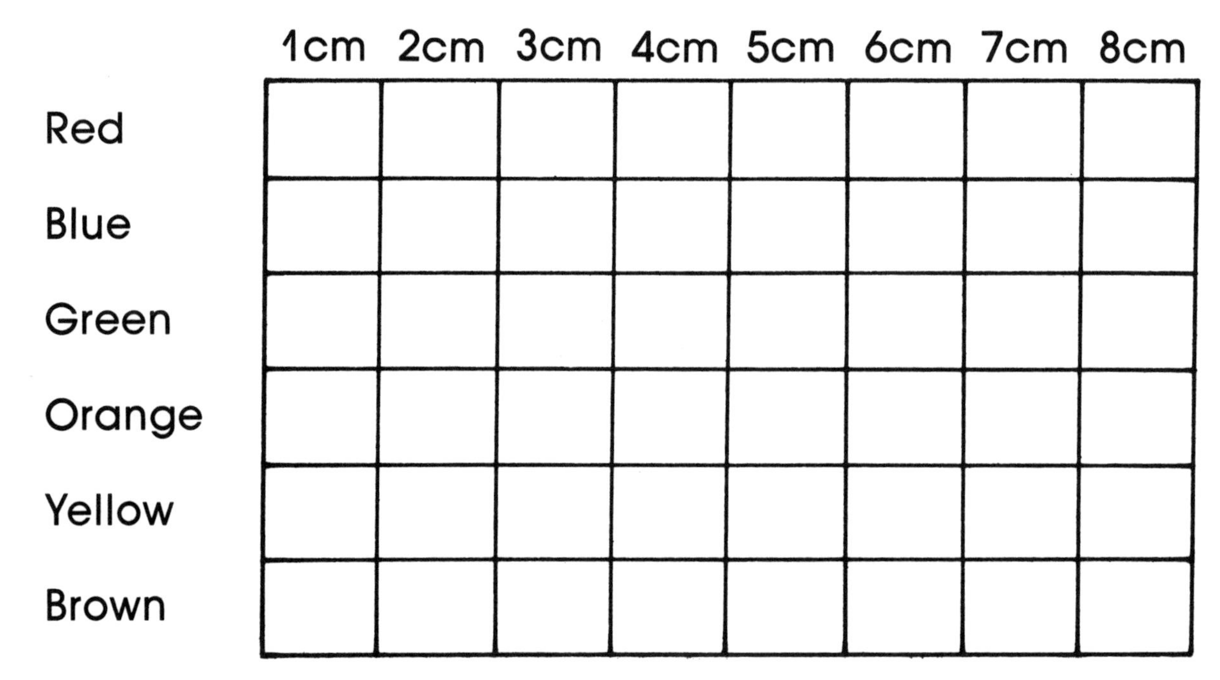

	1cm	2cm	3cm	4cm	5cm	6cm	7cm	8cm
Red								
Blue								
Green								
Orange								
Yellow								
Brown								

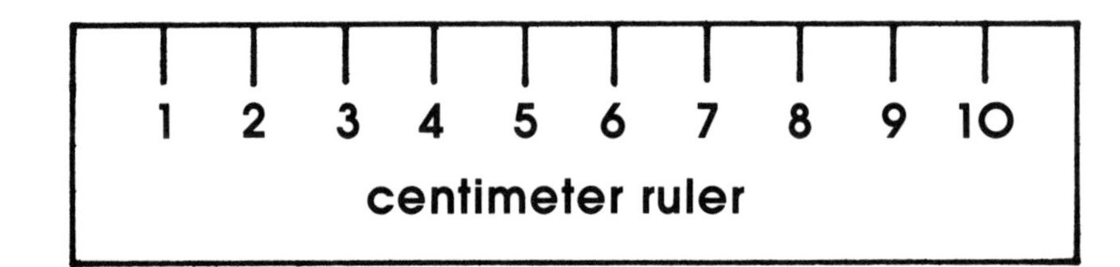

Skill: Reading a picture bar graph (counting and comparing)

Space Graph

Name ______________________________

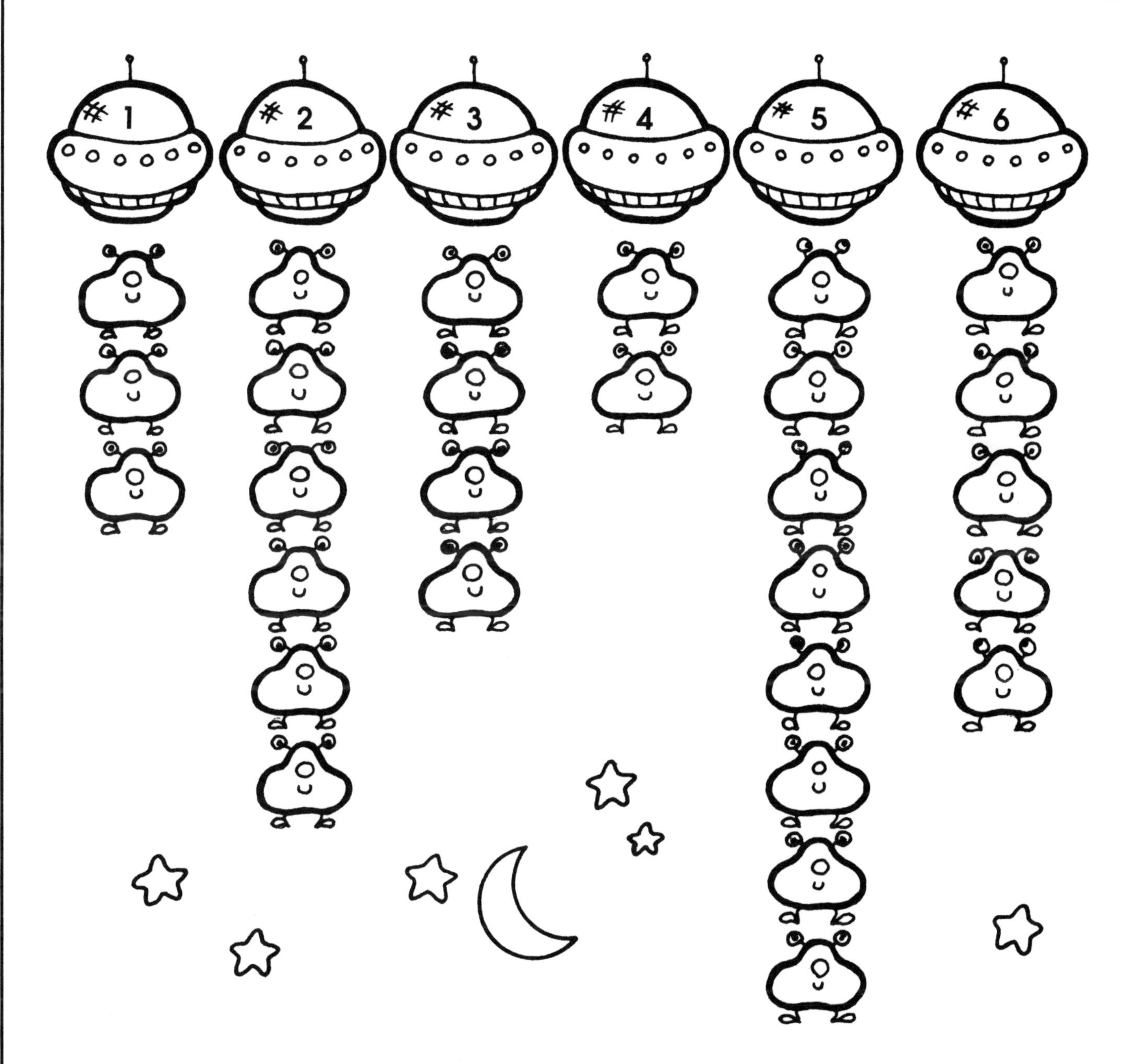

How many moonmen in spaceship 1?____ 2?____ 3?____ 4?____
5?____ 6?____

Which spaceship has the most moonmen? ____
Color them purple.

Which spaceship has the least moonmen? ____
Color them orange.

How many moonmen in all? ____

Skill: Reading a picture bar graph (counting and comparing)

Pickle Graph

Name ______________________________

Color the jars and pickles in the picture graph.

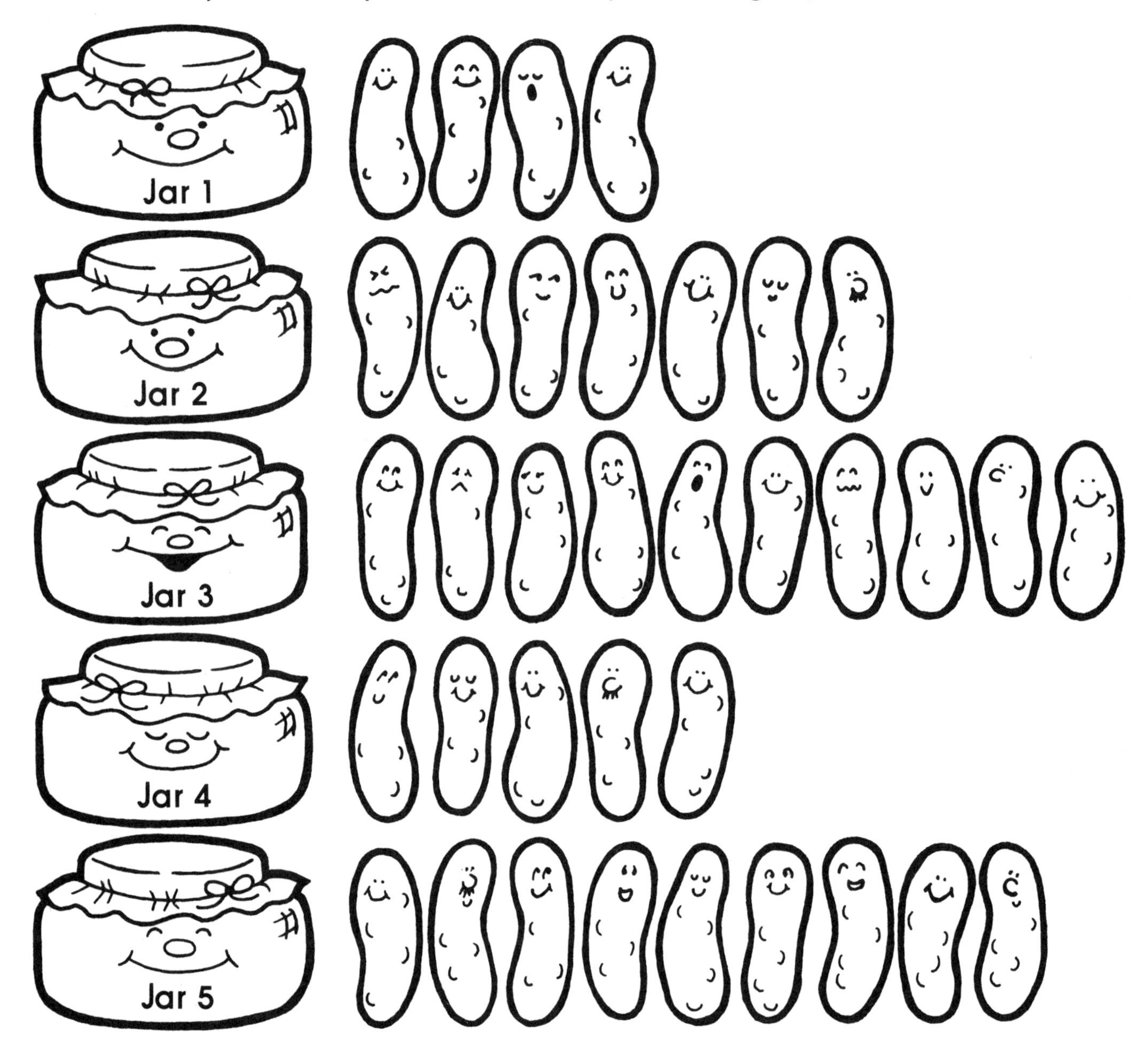

How many pickles?

Jar 1?____ Jar 2?____ Jar 3?____ Jar 4?____ Jar 5?____

In jars 2 and 3 together? ____ In jars 1 and 4 together? ____

How many more pickles in jar 3 than in jar 5? ____

In jar 2 than jar 1? ____ In jar 2 than jar 4? ____

In jar 4 than jar 1? ____

Skill: Reading a picture bar graph (counting)

Pat's Fish

Name ______________________

This picture graph shows how many fish Pat caught.

Color the fish Pat caught on the third Saturday red.
Color the fish he caught on the first Saturday blue,
the second Saturday yellow, and the fourth Saturday green.
How many fish did he catch on the first Saturday? ____
second Saturday? ____ third Saturday? ____ fourth Saturday? ____

Be Sharp!

Name ______________________________

1 2 3 4 5 6 7 8 9 10 11 12 13 14 15 16 17 18

The pencil above is about 11 centimeters long. Use your centimeter ruler to measure the other pencils.

about _____ centimeters

about _____ centimeters

about _____ centimeters

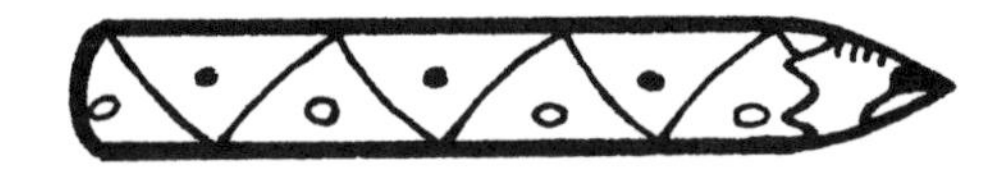

about _____ centimeters

about _____ centimeters

about _____ centimeters

about _____ centimeters

about _____ centimeters

about _____ centimeters

Skill: Measuring centimeters

You Measure Up!

Name ______________________________

The picture of the girl is about 8 centimeters high. How many centimeters high are these lines? Write your answers in the blanks.

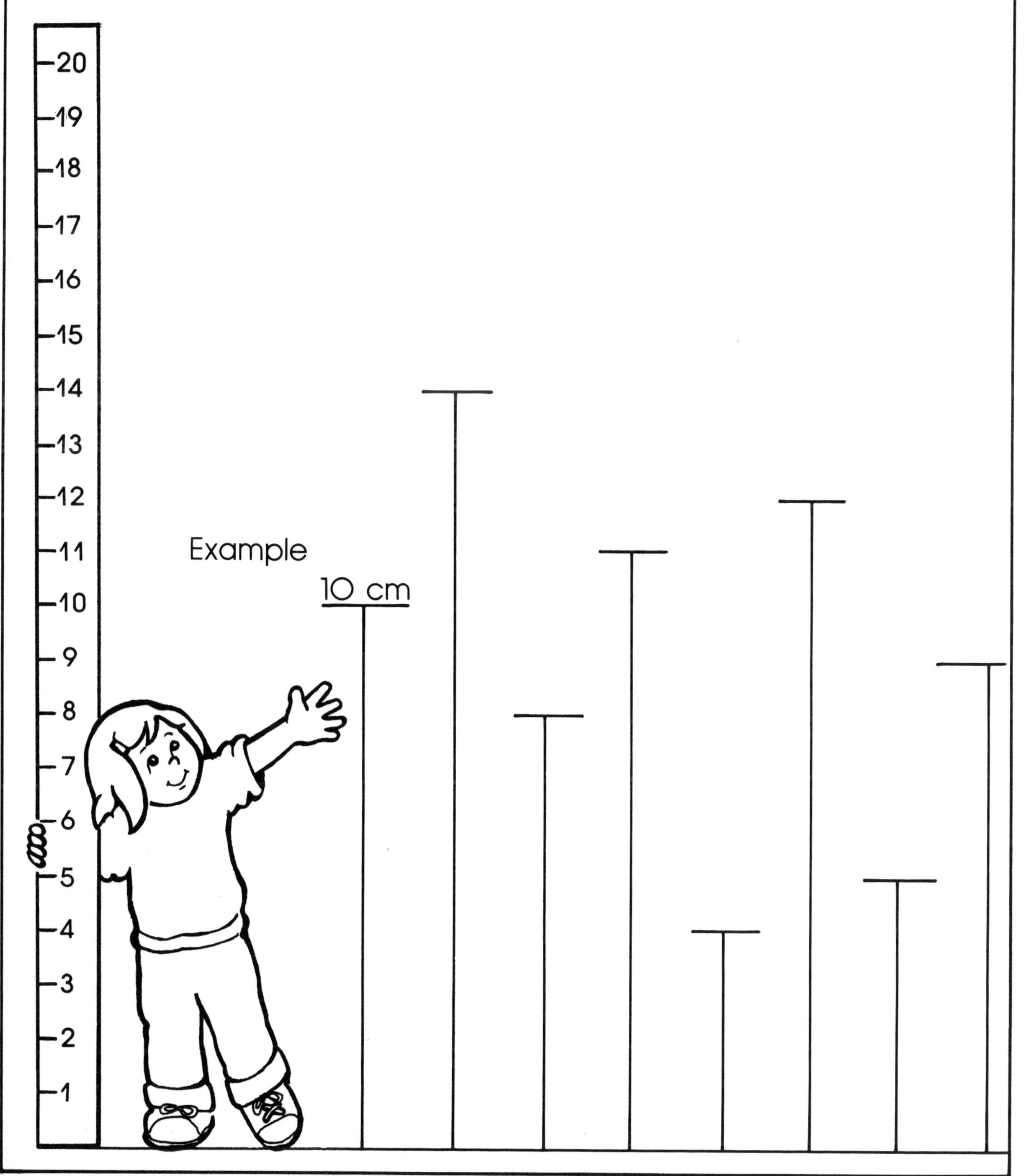

Brush Up on Measuring!

Name ______________________________

Use your centimeter ruler to measure these brushes to the nearest centimeter.

about _____ centimeters

about _____ centimeters

about _____ centimeters

about _____ centimeters

about _____ centimeters

about _____ centimeters

about _____ centimeters

about _____ centimeters

about _____ centimeters

about _____ centimeters

Skill: Measuring to the nearest inch

The Inch Worm

Name ________________________

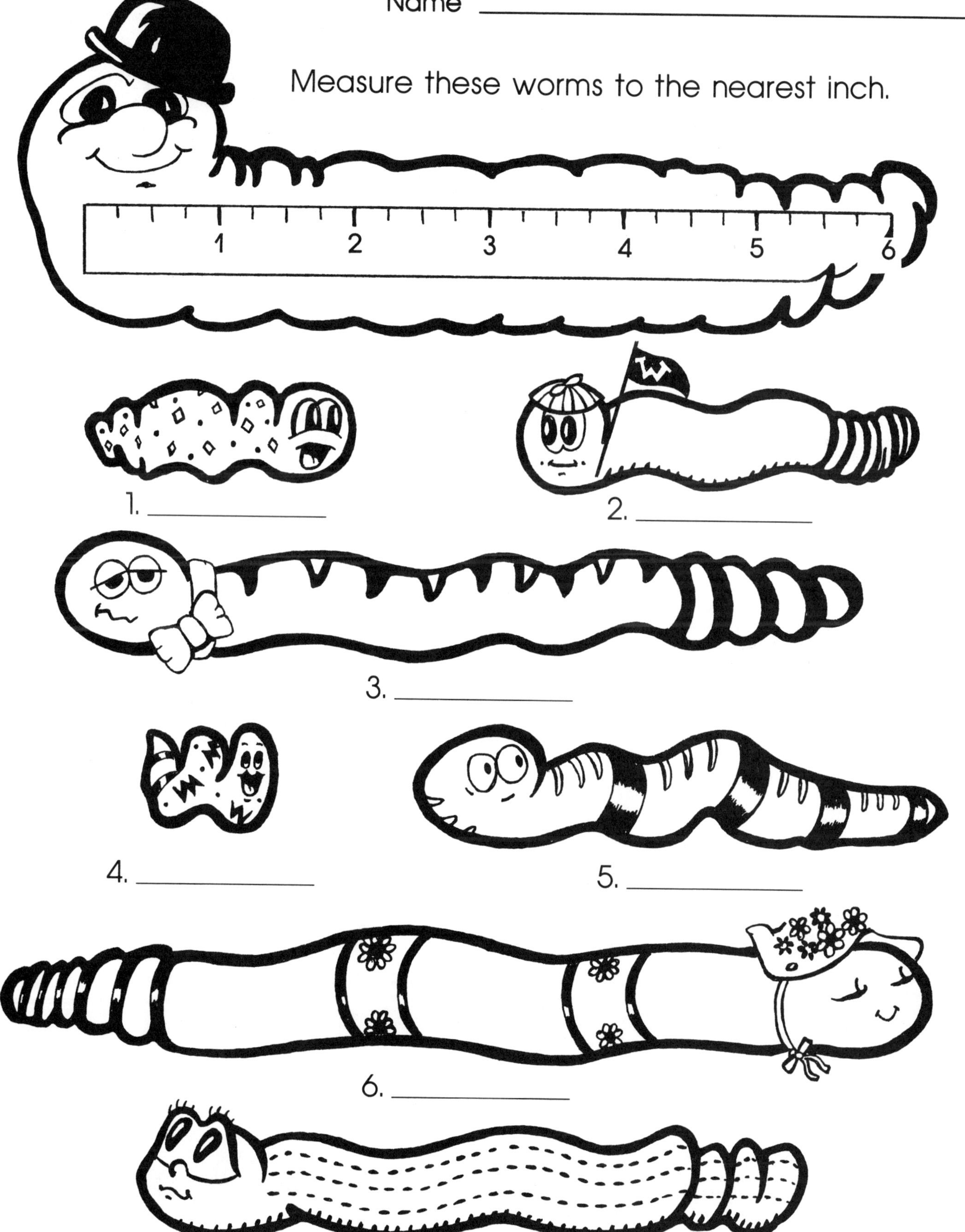

Hot Dog!

Name ________________________

How long are these hot dogs? Measure them to the nearest inch.

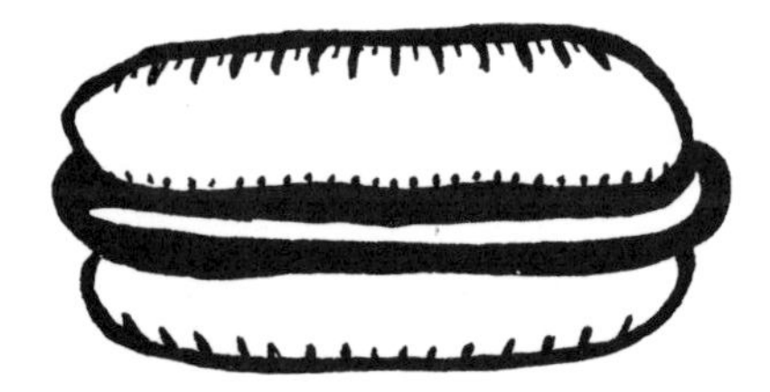

1. about ____ inches

2. about ____ inches

3. about ____ inches

4. about ____ inch

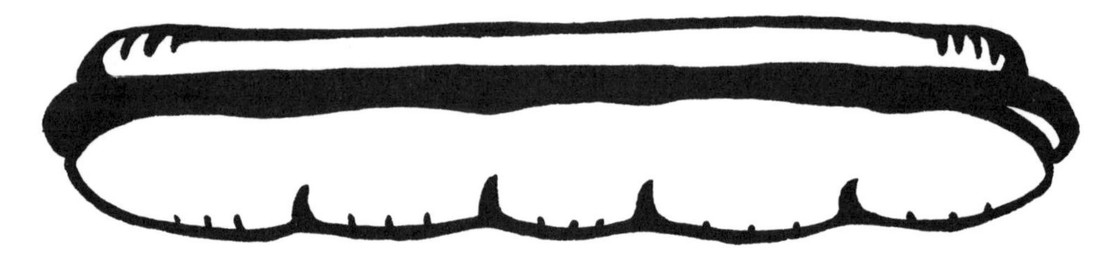

5. about ____ inches

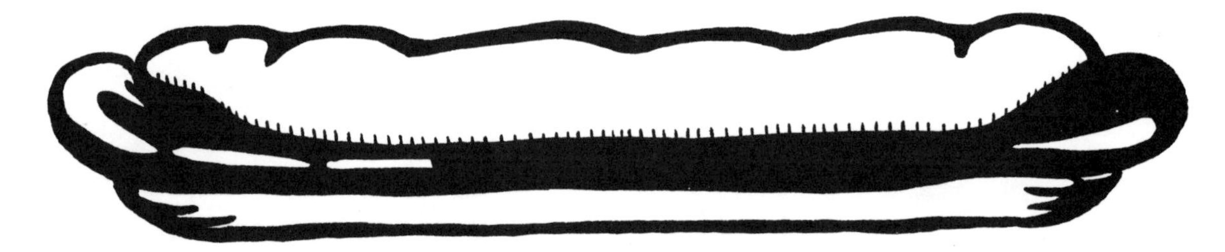

6. about ____ inches

7. about ____ inches

Measure for Treasure

Name ______________________

Ahoy, Matey! Go on a treasure hunt and measure to the nearest inch. Use the *'s to help you measure.

How far is it . . .

1. from the shark to the west edge of Treasure Island? __________
2. from the north edge of Skull Island to the shark? __________
3. from the east edge of the pirate ship to Pirate Cove? __________
4. from the north edge of Skull Island to the whale? __________
5. from the south edge of Treasure Island to the north edge of Pirate Cove? __________
6. from the letter in the bottle to the whale? __________
7. from the pirate ship to the treasure chest? __________

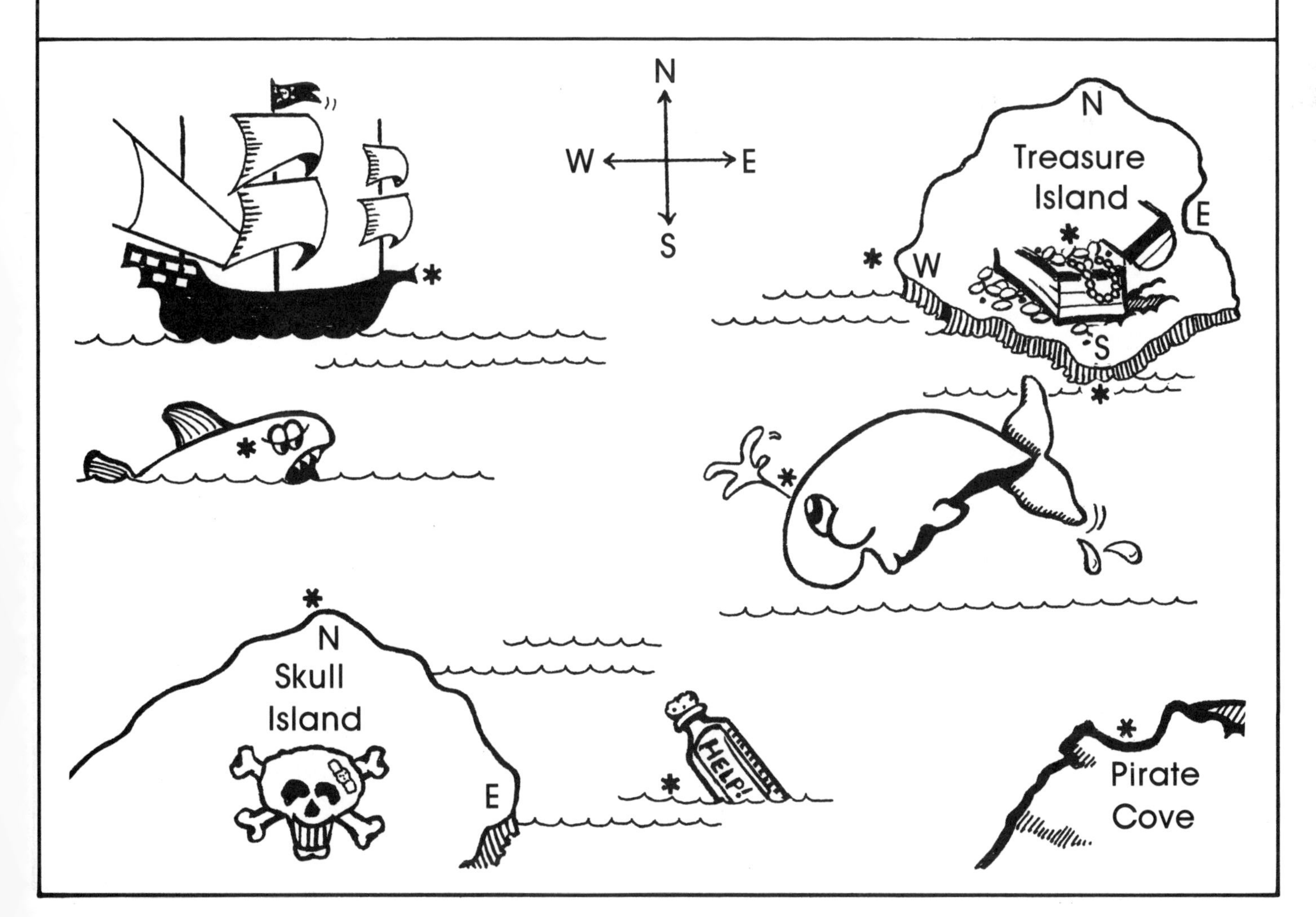

Block Heads

Name ____________________

How far is it around each block head. Count the centimeters. The first one is done for you.

Example

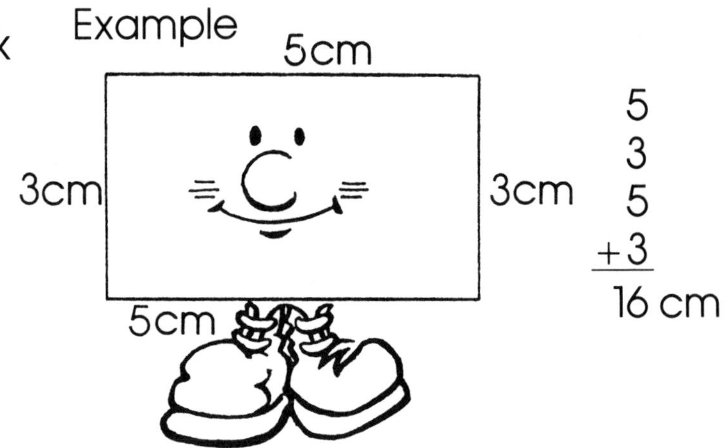

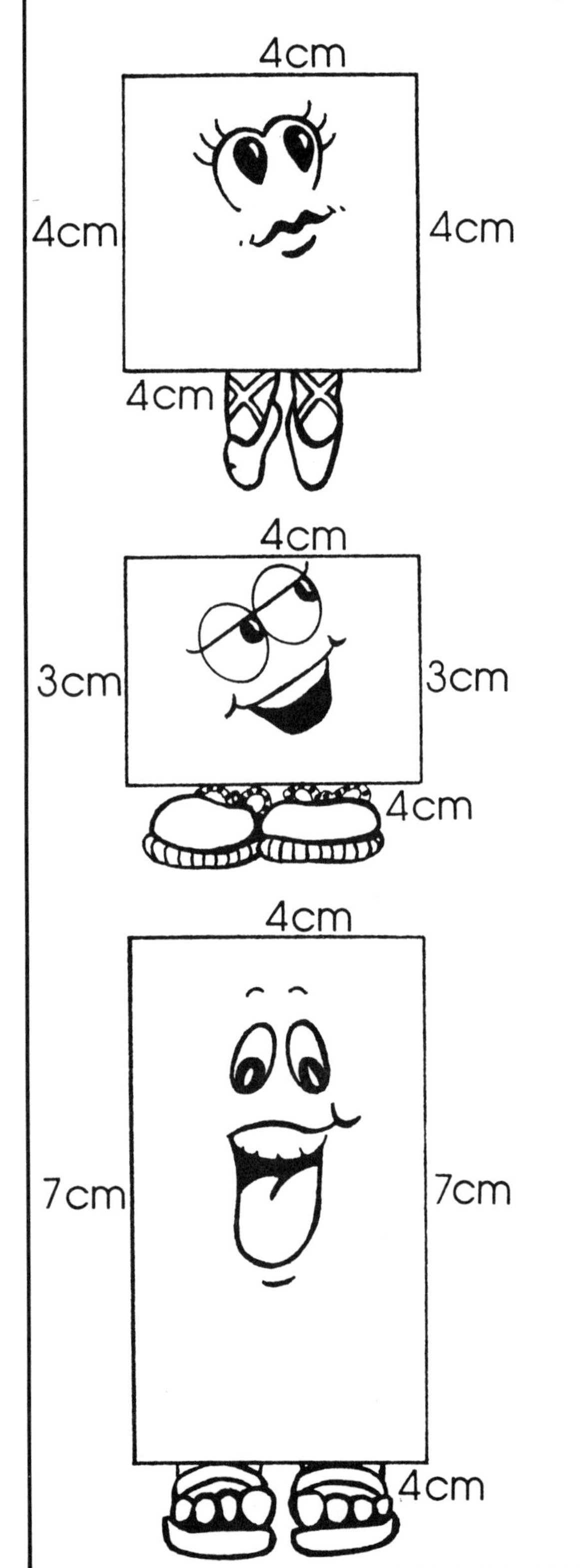

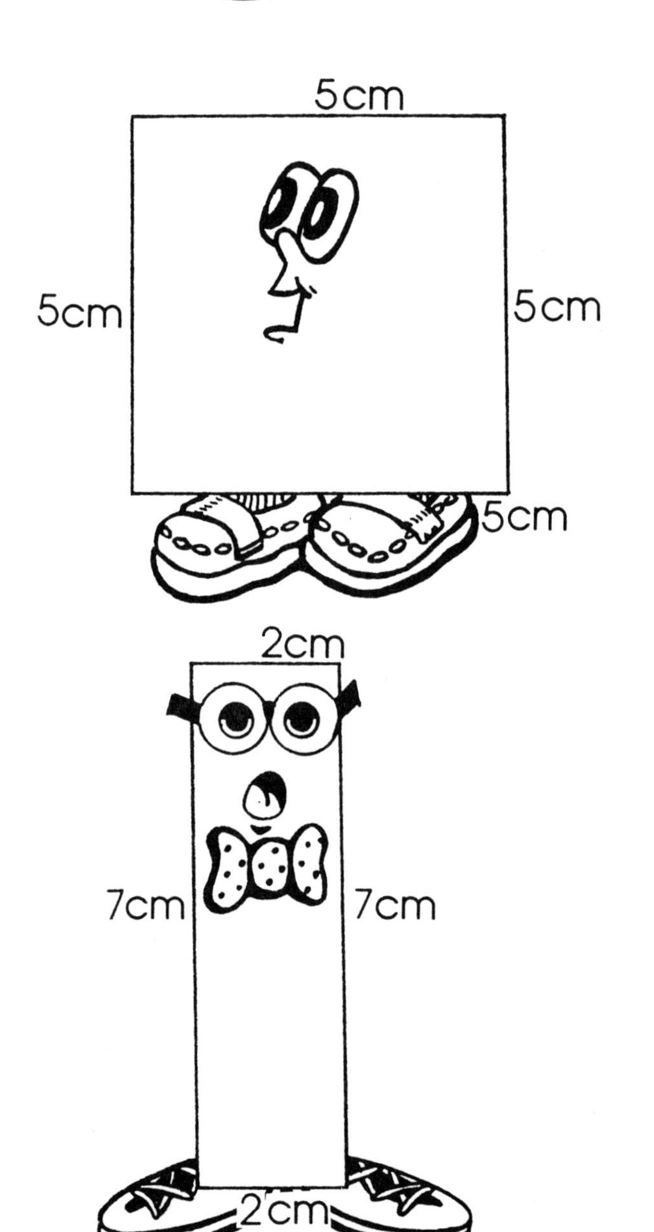

Skill: Measuring and drawing to the nearest inch

Tape Measures

Name ______________________________

Using an inch ruler, draw "tape" lines for the lengths given.

2 inches

5 inches

4 inches

3 inches

6 inches

1 inch

5 inches

4 inches

Cubs and Cubes

Name ______________________________

Help these cubs count the cubes in the boxes. Write the answer in the bear.

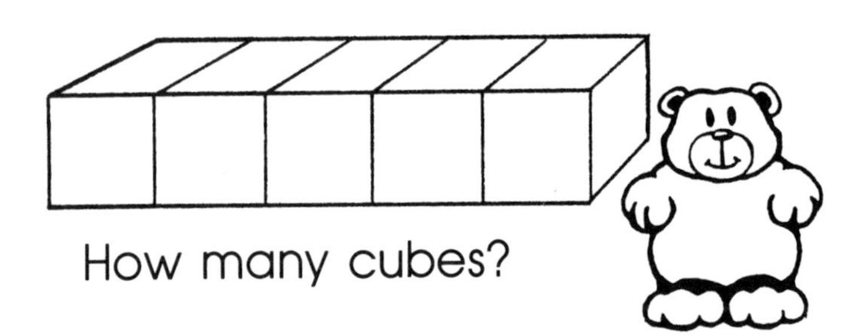
How many cubes?

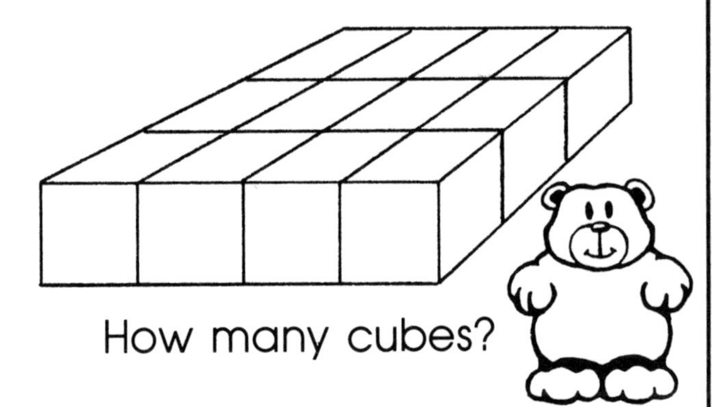
How many cubes?

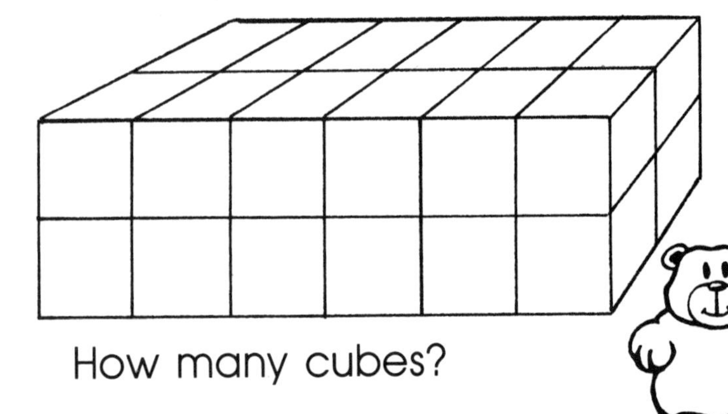
How many cubes?

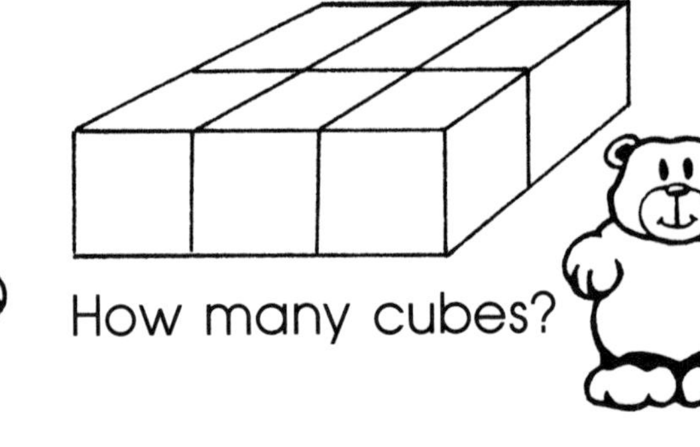
How many cubes?

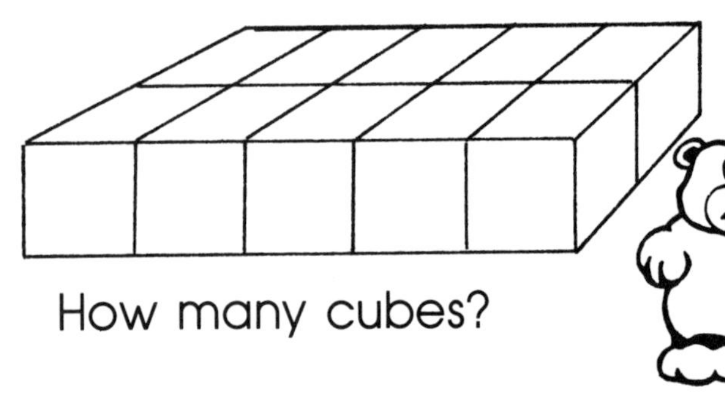
How many cubes?

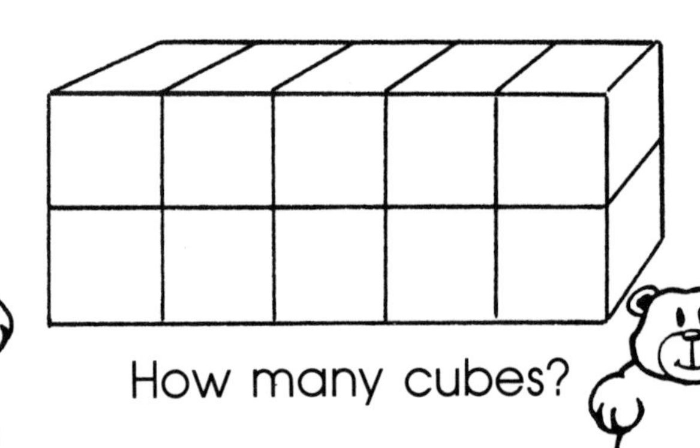
How many cubes?

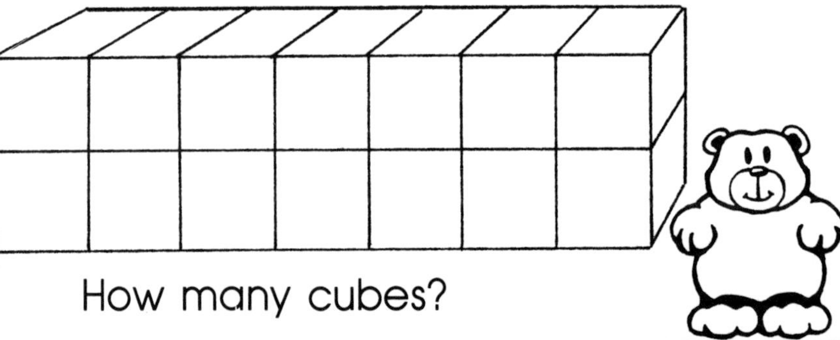
How many cubes?

Square Fair

Name ______________________________

How many squares are in each shape? These are called **square inches**.

inch

_____ square inches

inch

_____ square inches

How many squares are in each shape? These are called **square centimeters**.

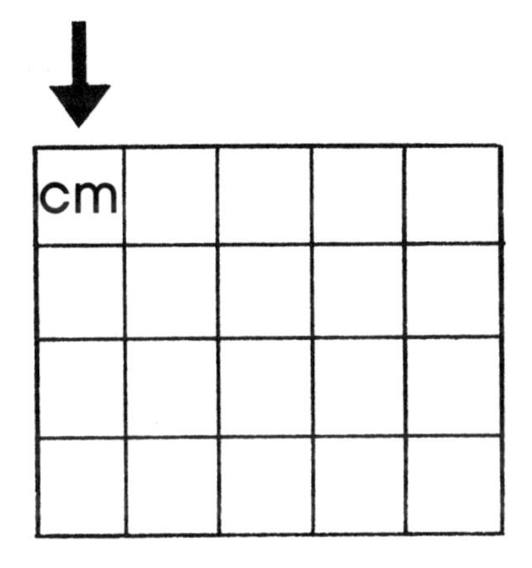

_____ square centimeters

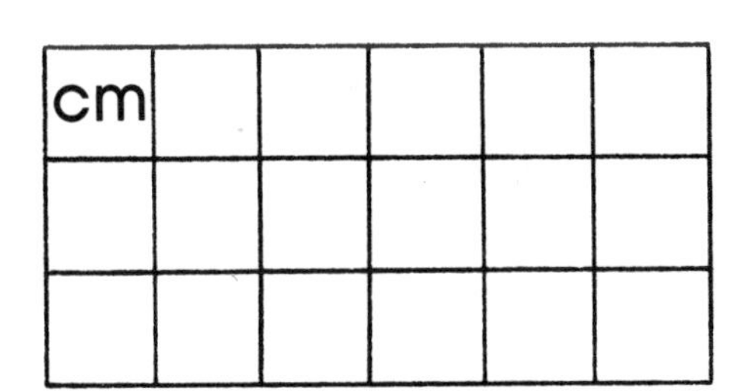

_____ square centimeters

cm

_____ square centimeters

Skill: Determining equivalent measures of volume – prints and quarts

Liquid Limits

Name ____________________

Draw a line from the containers on the left to the containers on the right which will hold the same amount of liquid.

Remember:
2 pints = 1 quart

Name

Picture Problems – I

Circle the picture which matches the number sentence.

1. $1 + 2 = 3$

or

2. $2 + 3 = 5$

or

3. $4 + 2 = 6$

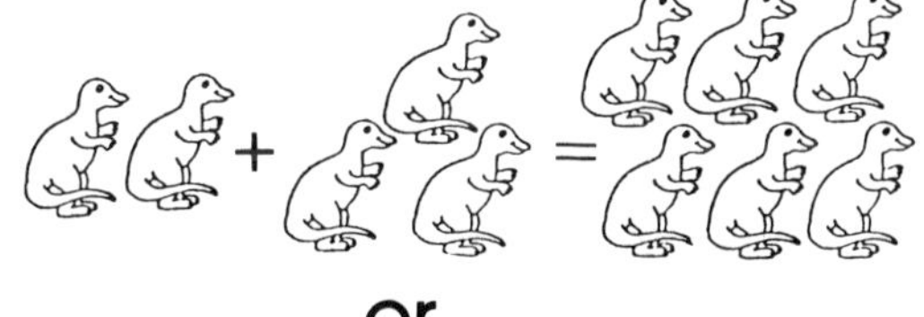

or

4. $5 + 1 = 6$

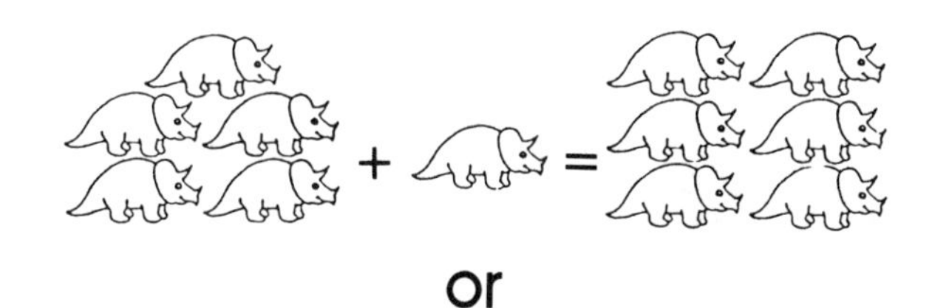

or

5. $3 + 4 = 7$

or

6. $6 + 1 = 7$

or

Name ______________________

Picture Problems – II

Look at the pictures and finish the number sentences.

1.

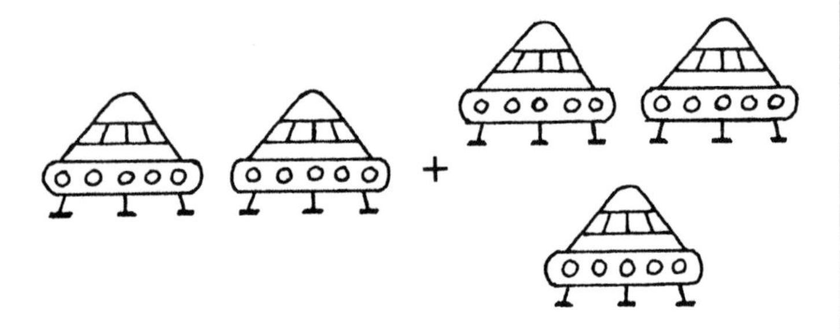

$2 + 3 =$ 5

2.

$1 + 7 =$ ______

3.

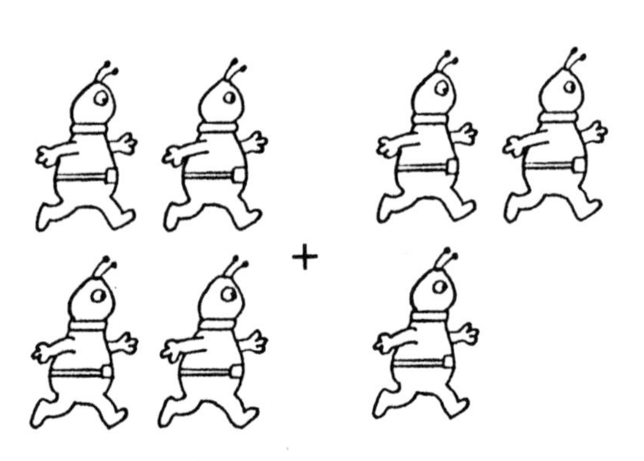

$4 + 3 =$ ______

4.

$5 + 0 =$ ______

5.

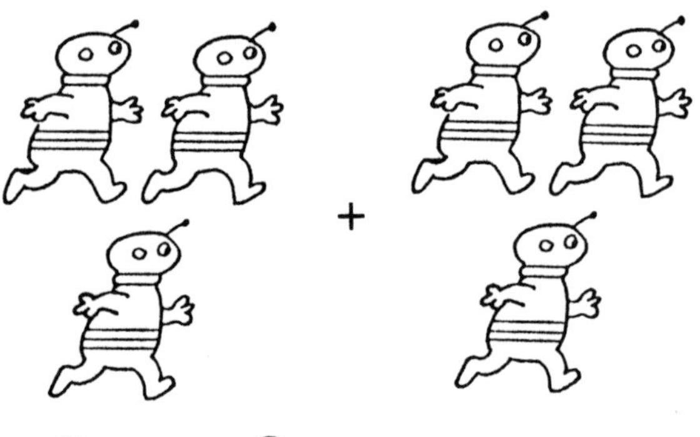

$3 + 3 =$ ______

6.

$4 + 5 =$ ______

Name ______________________

How Many More Are Needed?

Draw in the missing pictures and finish the number sentences.

1.

 + =  

1 + 2 = 3

2.

 + = 

3 + ____ = 5

3.

 + =

5 + ____ = 8

4.

 + = 

3 + ____ = 6

5.

 + =

2 + ____ = 7

6.

 + = 

4 + ____ = 5

Skill: Using additive subtraction

Name ______________________

How Many More Are Needed? – Review

Draw in the missing pictures and finish the number sentences.

1.

\+ =

1 + 2 = 3

2.

\+ =

____ + 3 = 6

3.

\+ =

5 + ____ = 7

4.

\+ =

____ + 3 = 5

5.

\+ =

____ + 4 = 8

6.

\+ =

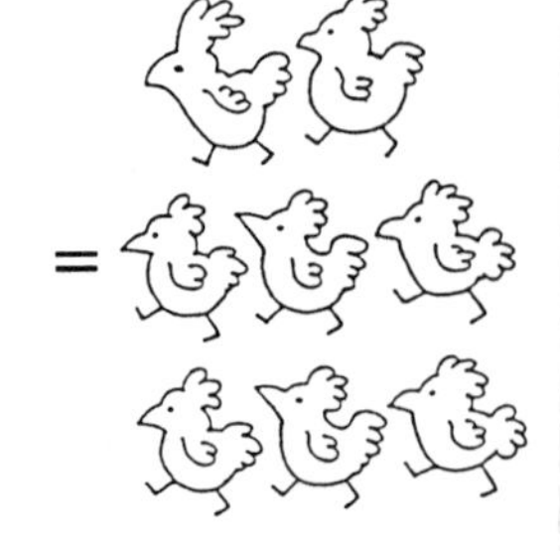

7 + ____ = 8

Name ____________________

Picture Problems – III

Circle the picture which matches the number sentence. Then finish the number sentence.

1.

or

$4 - 1 = 3$

2.

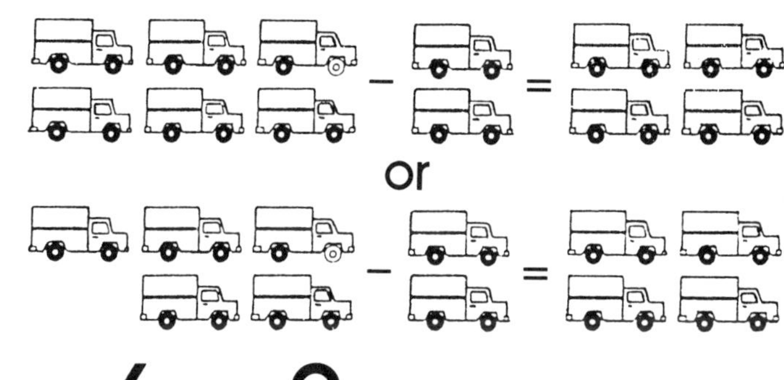

$6 - 2 =$ ______

3.

or

$5 - 3 =$ ______

4.

or

$7 - 3 =$ ______

5.

or

$5 - 2 =$ ______

6.

or

$7 - 5 =$ ______

Skill: Introducing the addition clue "in all"

Name ____________________

How Many in All? – I

Look at the pictures and finish the number sentences.

1.

 + 

How many 's are there in all?

$2 + 4 =$ 6

2.

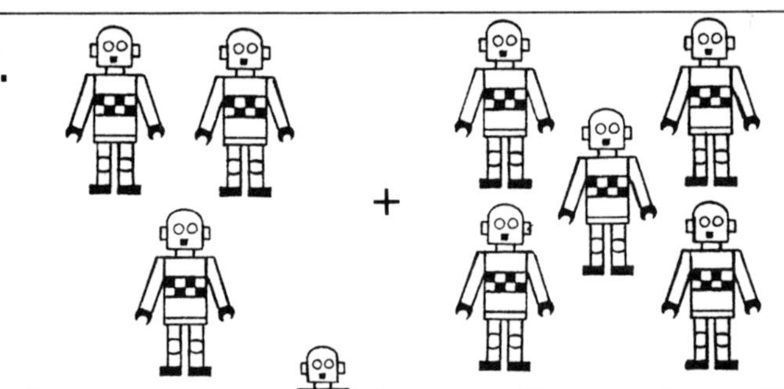

How many 's are there in all?

$3 + 5 =$ ______

3.

 + 

How many 's are there in all?

$4 + 3 =$ ______

4.

 + 

How many 's are there in all?

$4 + 1 =$ ______

5.

  + 

How many 's are there in all?

$2 + 5 =$ ______

6.

    +  

How many 's are there in all?

$4 + 4 =$ ______

Name ______________________

How Many in All? – II

Look at the pictures and finish the number sentences.

1.

How many 's are there in all?

$1 + 1 =$ 2

2.

How many 's are there in all?

$3 + 6 =$ ______

3.

How many 's are there in all?

$6 + 1 =$ ______

4.

How many 's are there in all?

$3 + 4 =$ ______

5.

How many 's are there in all?

$4 + 5 =$ ______

6.

How many 's are there in all?

$2 + 3 =$ ______

Name ____________________

How Many Are Left? – I

Look at the pictures and finish the number sentences.

1.

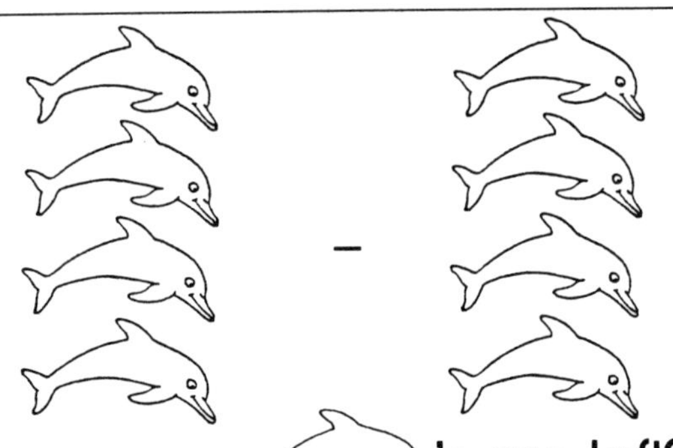

How many 's are left?

$4 - 4 =$ <u>0</u>

2.

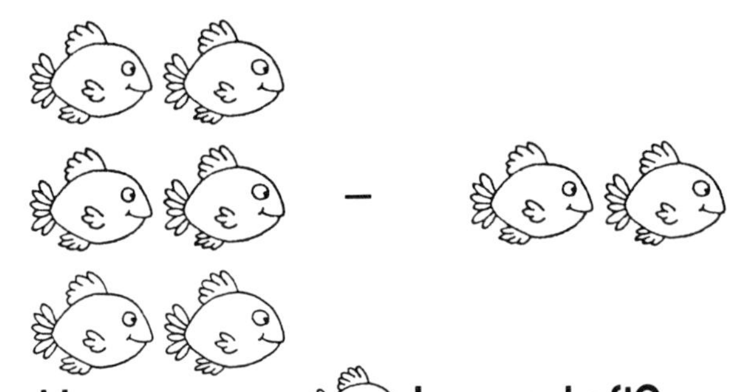

How many 's are left?

$6 - 2 =$ ______

3.

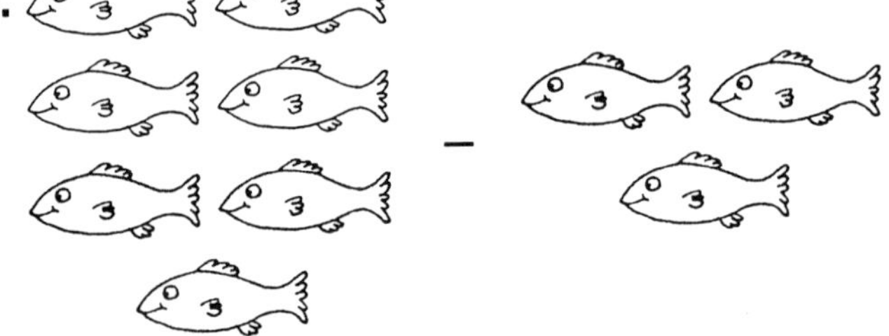

How many 's are left?

$7 - 3 =$ ______

4.

How many 's are left?

$6 - 5 =$ ______

5.

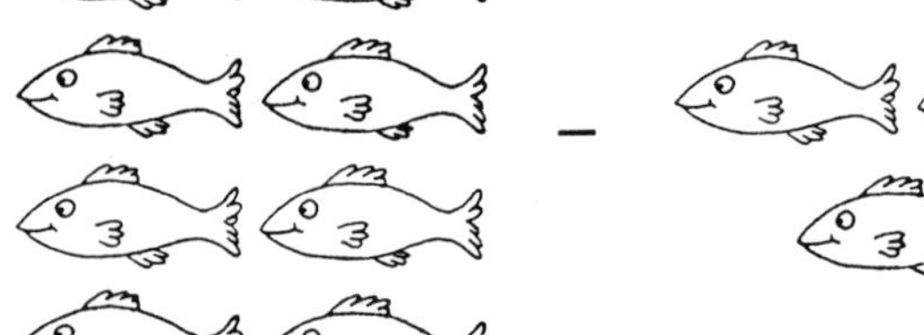

How many 's are left?

$8 - 3 =$ ______

6.

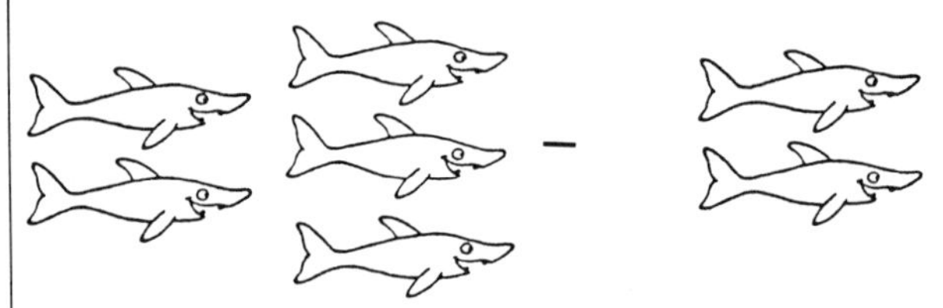

How many 's are left?

$5 - 2 =$ ______

Skill: Discriminating between addition and subtraction problems

Name ____________________

Add or Subtract? – I

Look at the pictures and finish the number sentences.

1.

5 ○ 6 = 11

2.

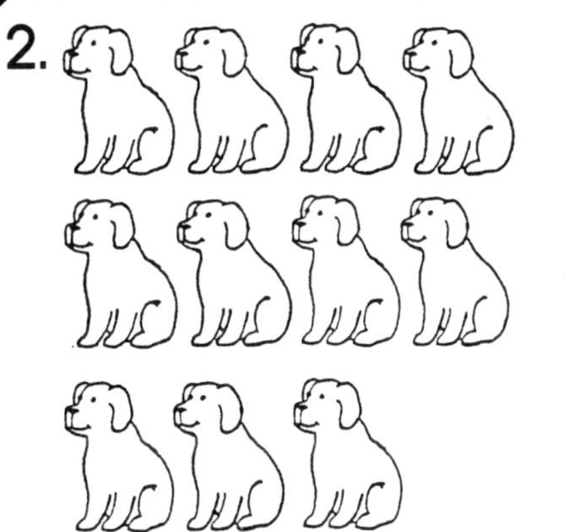

11 ○ 4 = ______

3.

12 ○ 7 = ______

4.

7 ○ 6 = ______

5.

5 ○ 5 = ______

6.

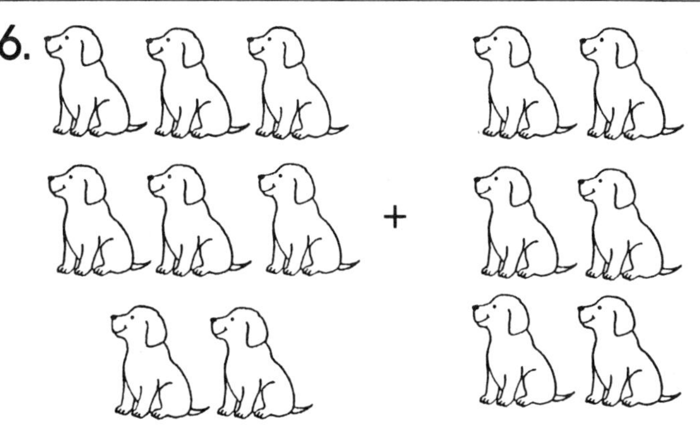

8 ○ 6 = ______

Skill: Discriminating between addition and subtraction problems

Name ____________________

Add or Subtract? – II

Look at the pictures and finish the number sentences.

1.

How many 's are there in all?

$7 \oplus 4 = \underline{11}$

2.

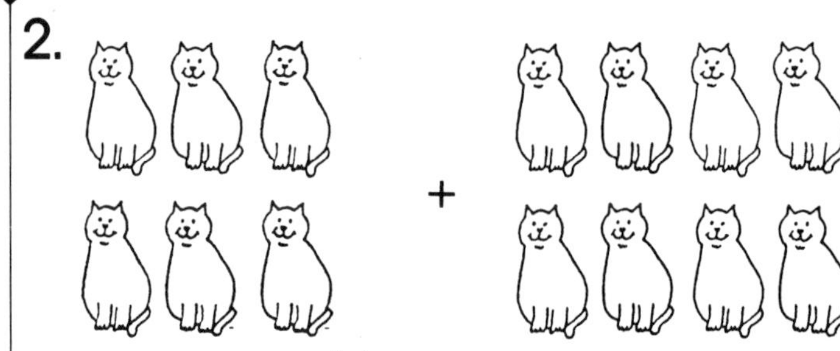

How many 's are there in all?

6 ○ 8 = ______

3.

How many 's are there in all?

11 ○ 2 = ______

4.

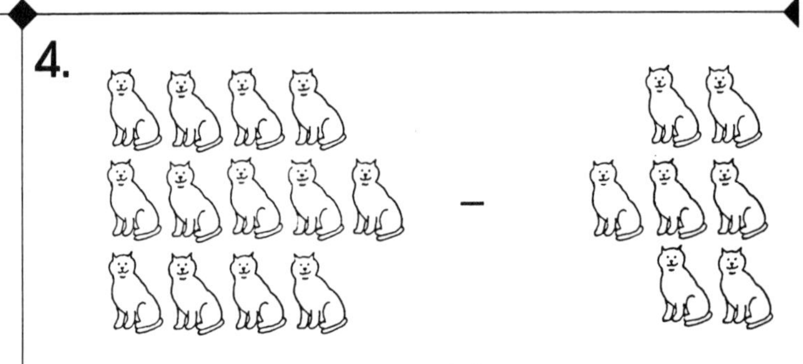

How many 's are left?

13 ○ 7 = ______

5.

How many 's are left?

9 ○ 6 = ______

6.

How many 's are left?

12 ○ 8 = ______

Name ____________________

What Was the Question?

Draw a line under the question that matches the picture. Then finish the number sentence.

1.

How many 's are there in all?

How many 's are left?

$11 - 7 =$ 4

2.

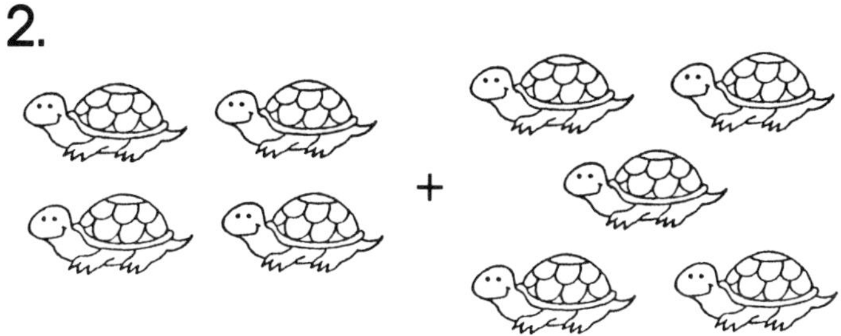

How many 's are there in all?

How many 's are left?

$4 + 5 =$ ______

3.

How many 's are there in all?

How many 's are left?

$8 - 3 =$ ______

4.

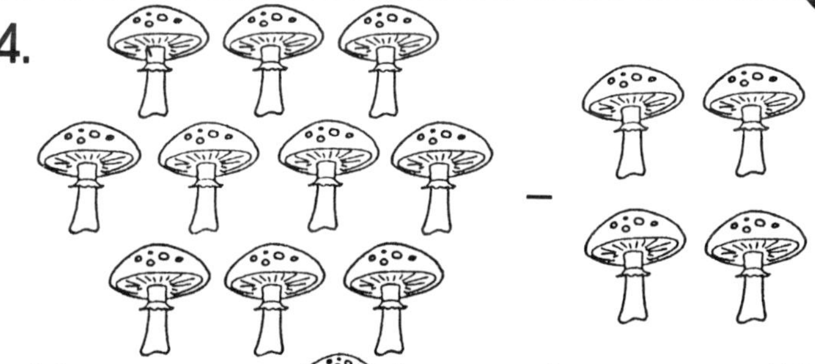

How many 's are there in all?

How many 's are left?

$10 - 4 =$ ______

5.

How many 's are there in all?

How many 's are left?

$5 + 6 =$ ______

6.

How many 's are there in all?

How many 's are left?

$8 + 4 =$ ______

Name ________________

What Was the Question? – Review

Draw a line under the question that matches the picture.
Then finish the number sentence.

1.

How many 's are there in all?
How many 's are left?

$6 + 6 = \underline{12}$

2.

How many 's are there in all?
How many 's are left?

$13 - 4 = \_\_\_\_$

3.

How many 's are there in all?
How many 's are left?

$9 + 5 = \_\_\_\_$

4.

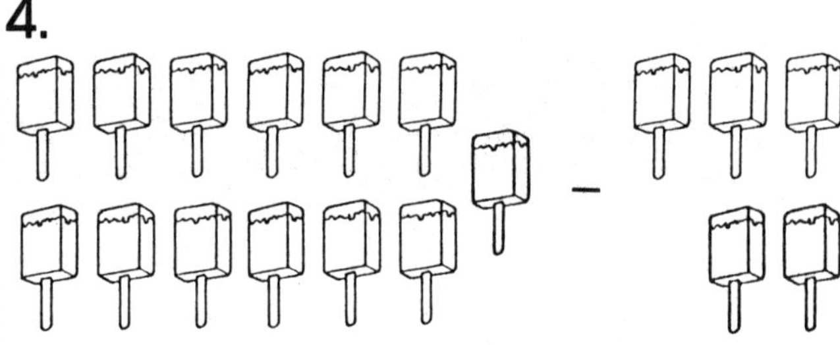

How many 's are there in all?
How many 's are left?

$13 - 5 = \_\_\_\_$

5.

How many 's are there in all?
How many 's are left?

$7 + 7 = \_\_\_\_$

6.

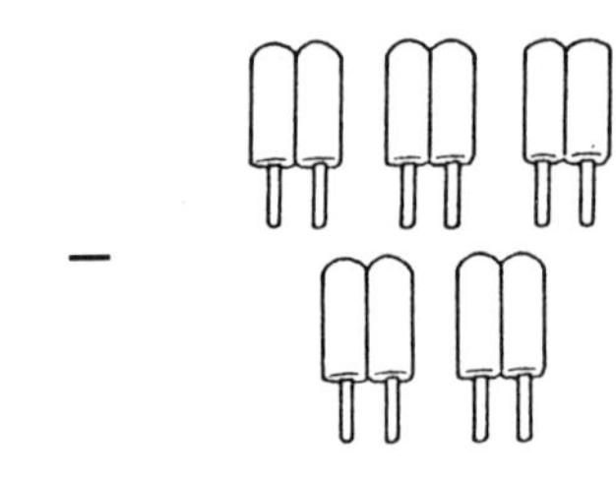

How many 's are there in all?
How many 's are left?

$9 - 5 = \_\_\_\_$

Name ____________________

Who Has the Most?

Circle the right answer.

1.
Jane has 3 's.
Bob has 4 's.
Bill has 5 's.

Who has the most 's?

Jane Bob Bill

2.
Pam has 7 's.
Joe has 5 's.
Jane has 6 's.

Who has the most 's?

Pam Joe Jane

3.
Amy has 23 's.
Sandy has 19 's.
Jack has 25 's.
Who has the most 's?

Amy Sandy Jack

4.
Ann has 19 's.
Burt has 18 's.
Brent has 17 's.
Who has the most 's?

Ann Burt Brent

5.
The boys have 14 's.
The girls have 16 's.
The teachers have 17 's.

Who has the most 's?

boys girls teachers

6.
Rose has 12 's.
Betsy has 11 's.
Ann has 13 's.
Who has the most 's?

Rose Betsy Ann

Skill: Using the key word "least"

Name

Who Has the Least?

Circle the right answer.

1.
Pat had 4 's.
Charles had 3 's.
Jane had 5 's.

Who had the least number of 's?

Pat Charles Jane

2.
Jeff has 5 's.
John has 4 's.
Bill has 6 's.

Who has the least number of 's?

Jeff John Bill

3.
Jane has 7 's.
Peg has 9 's.
Fred has 8 's.

Who has the least number of 's?

Jane Peg Fred

4.
Charles bought 12 's.
Rose bought 6 's.
Mother bought 24 's.

Who bought the least number of 's?

Charles Rose Mother

5.
John had 9 's.
Jack had 8 's.
Jeff had 7 's.

Who had the least number of 's?

John Jack Jeff

6.
Alma bought 12 's.
Nina bought 16 's.
Marty bought 13 's.

Who bought the least number of 's?

Alma Nina Marty

Name ______________________

How Many in All? III

The key words **in all** tell you to add. Circle the key words **in all** and solve the problems.

1. Jack has 4 white shirts and 2 yellow shirts. How many shirts does Jack have in all?

4 ⊕ 2 = ______

2. Joan has 4 pink blouses and 6 red ones. How many blouses does Joan have in all?

4 ◯ 6 = ______

3. Mack has 3 pairs of summer pants and 8 pairs of winter pants. How many pairs of pants does Mack have in all?

3 ◯ 8 = ______

4. Betsy has 2 black skirts and 7 blue skirts. In all, how many skirts does Betsy have?

2 ◯ 7 = ______

5. Willis has 5 knit hats and 5 cloth hats. How many hats does Willis have in all?

5 ◯ 5 = ______

Name ____________________

How Many in All? IV

Circle the addition key words **in all** and solve the problems.

1. On the block where Cindy lives there are 7 brick houses and 5 stone houses. How many houses are there in all?

$$7 + 5 = \_\_\_\_$$

2. One block from Cindy's house there are 7 white houses and 4 gray houses. How many houses are there in all?

3. Near Cindy's house there are 3 grocery stores and 5 discount stores. How many stores are there in all?

4. Children live in 8 of the two-story houses, and children live in 2 of the one-story houses. How many houses in all have children living in them?

5. In Cindy's neighborhood 4 students are in high school and 9 are in elementary school. In all, how many children are in school?

Name

How Many in All? V

Circle the addition key words **in all** and solve the problems.

1. At the park there are 3 baseball games and 6 basketball games being played. How many games are being played in all?

2. In the park 9 mothers are pushing their babies in strollers, and 8 are carrying their babies in baskets. How many mothers in all have their babies with them in the park?

3. On one team there are 6 boys and 3 girls. How many team members are there in all?

4. At one time there were 8 men and 4 boys pitching horseshoes. In all, how many people were pitching horseshoes?

5. While playing basketball, 4 of the players were wearing gym shoes and 6 were not. How many basketball players were there in all?

Name

How Many Are Left? – II

The key word **left** tells you to subtract. Circle the key word **left** and solve the problems.

1. Bill had 10 kittens, but 4 of them ran away. How many kittens does he have left?

$10 - 4 =$ ______

2. There were 12 rabbits eating clover. Dogs chased 3 of them away. How many rabbits were left?

3. Bill saw 11 birds eating from the bird feeders in his back yard. A cat scared 7 of them away. How many birds were left at the feeders?

4. There were 14 frogs on the bank of the pond. Then 9 of them hopped into the water. How many frogs were left on the bank?

5. Bill counted 15 robins in his yard. Then 8 of the robins flew away. How many robins were left in the yard?

Name ___________________________

How Many Are Left? III

Circle the subtraction key word **left** and solve the problems.

1. In Maggy's classroom there are 12 girls. One day 4 of the girls went home with the flu. How many girls were left in school that day?

2. Maggy is in 10 different clubs. This week 5 of them will not meet. How many of Maggy's clubs are left to meet this week?

3. Maggy had 16 crayons. She broke 9 of them. How many crayons does Maggy have left?

4. There are 13 boys in Maggy's classroom. One morning 8 of the boys went to the gym. How many were left in the classroom?

5. One day 4 of the 13 boys were called in from the playground. How many of the boys were left on the playground?

Name ______________________

Add or Subtract? – III

The key words **in all** tell you to add. The key word **left** tells you to subtract. Circle the key words and solve the problems.

1. The pet store has 3 large dogs and 5 small dogs. How many dogs are there in all?

   3 ⊕ 5 = ______

2. The pet store had 9 parrots and then sold 4 of them. How many parrots does the pet store have left?

   9 ◯ 4 = ______

3. The pet store gave Linda's class 2 adult gerbils and 9 young ones. How many gerbils did Linda's class get in all?

   2 ◯ 9 = ______

4. At the pet store 3 of the 8 myna birds were sold. How many myna birds are left in the pet store?

   8 ◯ 3 = ______

5. The monkey at the pet store has 5 rubber toys and 4 wooden toys. How many toys does it have in all?

   5 ◯ 4 = ______

Name

Add or Subtract? – IV

Circle the key words **in all** and **left** and solve the problems.

1. Alex picked 5 tomatoes, and Don picked 4. How many tomatoes did they pick in all?

2. George picked 17 apples and ate 6 of them. How many apples does he have left?

3. Mary picked 4 peppers, and Sue picked 9 peppers. How many peppers in all did they pick?

4. There were 11 roses on the bush. Cindy picked 5 of them. How many roses were left on the bush?

5. Max picked 8 bunches of grapes. Joe picked 6 bunches. How many bunches of grapes were picked in all?

Name ______________________

Key Word Practice – I

The key word **altogether** tells you to add. They key words **many more** tell you to subtract. Circle the key words and solve the problems.

1. Judy has 9 pennies. Mack has 3 pennies. How many more pennies does Judy have than Mack?

   9 – 3 = ______

2. Sarah has 12 dimes. Joe has 7 dimes. How many more dimes does Sarah have than Joe?

3. Sally was paid 8 quarters for cutting the lawn. She already had 4 quarters. How many quarters altogether does Sally have?

4. Jim had 5 coins. His dad gave him 9 more. How many coins does Jim have altogether?

5. Beth earned 2 dollars. Martha earned 11 dollars. How many more dollars did Martha earn than Beth?

Name

Key Word Practice – II

Circle the key words **altogether** and **many more.** Then solve the problems.

1. Jill rode the merry-go-round 7 times yesterday and 6 times today. How many times did she go on the ride altogether?

2. Don rode the pony for 17 minutes. Dave's pony ride lasted 12 minutes. How many more minutes did Don ride than Dave?

3. Mary has 8 balloons. Tina has 14. How many balloons do they have altogether?

4. Jerry ate 10 hot dogs. Betty ate 7. How many more hot dogs were eaten by Jerry than by Betty?

5. Bob saw 11 shows at the fair. Marie saw 9 shows. How many shows did they see altogether?

Name ________________________

Fractions

Study the pictures. Then solve the story problems.

1. Bob ate 1/2 of a candy bar. Bill ate 1/4 of a candy bar. Who ate the most candy?

2. Sue ate 3/4 of a cupcake. Angie ate 1/2 of a cupcake. Who ate the most?

3. Pete ate 1/3 of a pie. May ate 1/4 of a pie. Who ate the most?

4. Emily ate 1/2 of an apple. Dean ate 3/4 of an apple. Who ate the most?

5. Amy sliced 1/4 of the cake. Sally sliced 1/3 of the cake. Who sliced the most cake?

6. John ate 1/2 of a melon. Jeff ate 2/3 of a melon. Who ate the most?

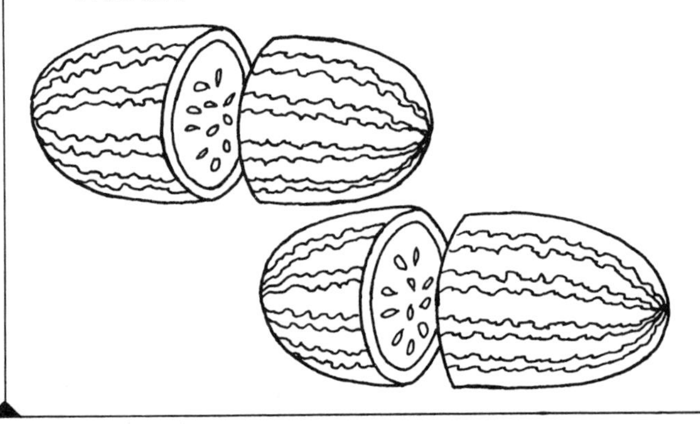

Skill: Discriminating between "most" and "fewer"

Name

Most or Fewer?

Solve the following story problems.

1. Margo swam 17 laps in the pool. Mark swam 23 laps. Who swam the most laps?

2. Jan hit 43 singles. Mack hit 39. Who hit the most singles?

3. Aaron stole 31 bases. Bob stole 29 bases. Who stole fewer bases?

4. Ellen's school won 37 events at the meet. Tom's school won 41 events. Whose school won fewer events?

5. Ellen's school has 129 athletes. Tom's school has 128 athletes. Whose school has the most athletes?

Name

Smallest, Largest, First, Last

In the series 3, 9, 5, 1, 7 the **smallest** number is 1. The **first** number in the series is 3. The **last** number is 7. The **largest** number is 9. Work the following problems.

1. Circle the first number.

17 20 12 5 6

2. Circle the last number.

17 20 12 5 6

3. Circle the largest number.

17 20 12 5 6

4. Circle the smallest number.

1 2 3 4 5 6

5. Circle the last number.

6 5 4 3 2 1

Skill: Discriminating between addition and subtraction problems

Name ________________

Which Number Sentence? – I

Circle the number sentence that matches the story problem. Then solve the problem.

1. There are 15 men and 8 women on the police force. How many police are there on the force in all?

$15 + 8 =$ ______

$15 - 8 =$ ______

2. The city council has 12 members. The school board has 4 fewer members than the city council. How many members does the school board have?

$12 + 4 =$ ______

$12 - 4 =$ ______

3. The school board hired 12 bus drivers and 2 mechanics. How many drivers and mechanics did they hire altogether?

$12 + 2 =$ ______

$12 - 2 =$ ______

4. The school board hired 47 teachers and 5 administrators. How many more teachers were hired than administrators?

$47 + 5 =$ ______

$47 - 5 =$ ______

5. The council bought 5 fire trucks and 4 garbage trucks. How many trucks in all were bought?

$5 + 4 =$ ______

$5 - 4 =$ ______

Name ______________________

Which Number Sentence? – II

Circle the number sentence that matches the story problem. Then solve the problem.

1. Last month 43 homeless adult dogs were kept at the pound. There were 11 fewer puppies than adult dogs at the pound. How many puppies were kept at the pound?

$43 + 11 =$ ______

$43 - 11 =$ ______

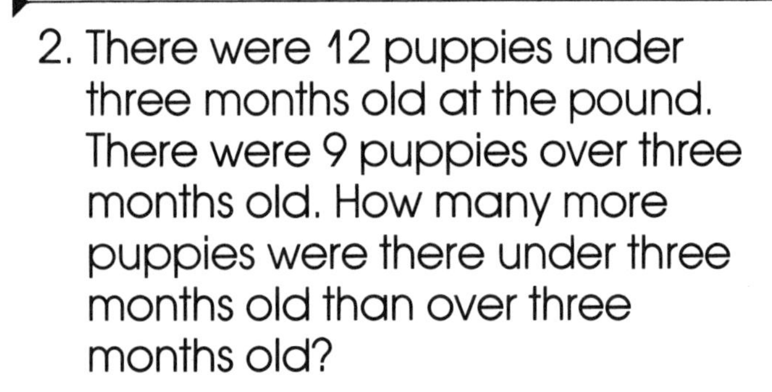

2. There were 12 puppies under three months old at the pound. There were 9 puppies over three months old. How many more puppies were there under three months old than over three months old?

$12 + 9 =$ ______

$12 - 9 =$ ______

3. Last week 8 black puppies and 7 brown puppies were given new homes. How many puppies in all were given new homes?

$8 + 7 =$ ______

$8 - 7 =$ ______

4. The puppies were fed 25 pounds of dog food on Monday and 9 pounds of dog food on Tuesday. How many more pounds of food did they eat on Monday than on Tuesday?

$25 + 9 =$ ______

$25 - 9 =$ ______

5. The dogs were given 18 gallons of water on Monday and 7 gallons of water on Tuesday. How many gallons did they drink on Monday and Tuesday altogether?

$18 + 7 =$ ______

$18 - 7 =$ ______

Skill: Simple Addition (number combinations to 18)

Wacky Waldo

Name ______________________

Wacky Waldo is a different kind of animal trainer. He teaches mice to chase cats, he shows snakes how to fly, and he teaches fish to walk on land. He even teaches turtles to run faster than rabbits.

1. Wacky Waldo taught 8 mice to chase cats. Then he taught 9 mice to chase cats. How many mice did he teach in all?

2. Wacky Waldo saw 11 cats chased by his mice on Saturday. On Sunday, he saw 6 cats chased. What was the sum?

3. Wacky Waldo taught 9 mice to scare elephants and 4 mice to chase lions. How many mice did he teach?

4. Waldo found 11 whiskers on a mouse named Fred. He found 4 whiskers on a cat named Kitty. What was the total?

5. On Monday, Waldo taught 6 snakes to fly. On Tuesday, he taught 7 snakes to fly. How many snakes learned to fly in all?

6. Waldo used to have 4 flying snakes. Then he got 14 more flying snakes. How many flying snakes does Waldo have now?

Ant Betty

Name ______________________

1. Ant Betty has 6 legs. Her friend Suzy Spider has 8 legs. How many more legs does Suzy have?

2. Ant Betty found 15 grass seeds. Mike Mouse took 6 of them. How many seeds are left?

3. Ant Betty has 18 brothers and 7 sisters. How many more brothers does she have?

4. Suzy Spider caught 13 flies last week. This week, she caught 5 flies. How many more did she catch last week?

5. Ant Betty ate 8 seeds this week. Last week, she ate 12 seeds. How many more did she eat last week?

6. Suzy Spider walked 17 feet in an hour. Ant Betty walked only 9 feet in an hour. How much farther did Suzy walk?

7. Ant Betty has 12 girl cousins. She has 17 boy cousins. How many more boy cousins does she have?

Ant Betty's Farm

Name ____________________

1. Ant Betty has 12 brothers and 10 sisters. How many brothers and sisters does she have in all?

2. Ant Betty picked up 15 seeds in the morning and 13 seeds in the afternoon. How many seeds did she pick up altogether?

3. Ant Betty found 17 bits of sugar. Her brother found 11 bits of sugar. How much sugar did they find in all?

4. Betty found 33 bread crumbs. Her friend Sue found 26 bread crumbs. How many did they find in all?

5. Betty has 14 girl cousins and 15 boy cousins. What is the total?

6. Ant Betty saw 23 flies in her garden and 13 flies in her house. What was the sum?

7. Ant Betty worked 42 hours one week. The next week, she worked 54 hours. What was the total?

To Tell the Tooth

Name ______________________________

1. Children have 20 teeth. Adults have 32 teeth. How many more teeth do adults have?

2. Possums have 50 teeth. Cats have 30 teeth. How many more teeth do possums have?

3. A wild pig has 44 teeth. An aardvark has 20 teeth. What is the difference?

4. A house mouse has 16 teeth. An anteater has 0 teeth. How many more teeth does the mouse have?

5. An elephant has 32 teeth. A sea lion has 36 teeth. What is the difference?

6. A hippopotamus has 36 teeth. A kangaroo has 34 teeth. What is the difference?

7. A mongoose has 40 teeth. A possum has 50 teeth. How many more teeth does the possum have?

Wacky Waldo

Name ____________________

Wacky Waldo is a great animal trainer. He teaches elephants to dance, rabbits to scare alligators and cows to fly like birds. He even teaches whales to water-ski and ride bicycles.

1. Waldo had 9 flying cows on Saturday and 11 flying cows on Sunday. How many flying cows did he have altogether?

2. Wacky Waldo taught 13 whales to water-ski and 27 whales to ride bicycles. How many whales did he teach in all?

3. Waldo's flying cows flew 16 miles at one time and 25 miles at another time. How many miles did they fly altogether?

4. Waldo taught Ricky Rabbit to scare alligators. Ricky scared 16 alligators the first day and 18 alligators the next day. How many alligators did he scare in all?

5. Waldo's dancing elephants did 16 shows on Monday and 19 shows on Tuesday. How many shows did they do altogether?

6. An elephant named Butterfly learned 15 dances in the morning and 6 dances in the afternoon. How many dances did she learn in all?

The Busy Tooth Fairy

Name ____________________

1. Adam Aardvark had 20 teeth. He lost 11 teeth. How many teeth did he have left?

2. Candy Kitten had 26 teeth. She now has 17 teeth. How many teeth did the busy tooth fairy pay for?

3. Harry Hare used to have 28 teeth. He still has 19 teeth. How many teeth did the tooth fairy pay for?

4. Kenny Kangaroo had 34 teeth. 15 teeth fell out. How many were left?

5. Peter Possum used to have 50 teeth. He lost 23 of them when he fell off the monkey bars. How many were left?

6. Sam Seal lost 19 teeth. He used to have 34 teeth. How many teeth does he have now?

7. Henry Hippo had 36 teeth. He lost 17 teeth. How many teeth does he have left?

Smelly Belly's Picnic

Name ____________________

1. Smelly Belly invited her friends to a picnic. She brought 17 bees, and her friend Stinky brought 14 bees. What was the total?

2. Stinky brought 17 eggs to the picnic. Smelly brought 14 eggs. How many were there altogether?

3. Smelly found 24 snails for the picnic. Only 18 were eaten. How many were left?

4. Smelly loves to eat wasps. She brought 54 fried wasps. She ate 27 of them herself. How many wasps were left?

5. Stinky came to the picnic with 43 crickets. Smelly came with 18 crickets. What was the total?

6. Smelly Belly came to the picnic with 45 wild roots. The other skunks at the picnic ate 26 of them. How many roots were left?

7. Stinky's favorite food is spiders. There were 32 at the picnic. He ate 29 spiders. How many were left?

Lizzy the Lizard — Bug Collector

Name ______________________________

Lizzy the Lizard loves to collect bugs. She has a collection of beetles, crickets and flies. She would have more, but sometimes she forgets and eats part of her collection.

1. Lizzy caught 18 beetles and 21 flies. How many more flies did she catch?

2. Lizzy found 19 bugs in a garden. Her friend Dizzy found 17. How many did they find in all?

3. Lizzy had 31 flies on Sunday. She forgot and ate 12 of them. How many were left?

4. Lizzy chased 34 butterflies. She caught 25 of them. How many got away?

5. Lizzy found 12 beetles on Monday and 19 beetles on Tuesday. How many did she find in all?

6. Lizzy caught 15 moths and 17 stinkbugs. How many insects did she catch altogether?

Worm Burgers

Name ______________________________

Worm Burgers is a new fast-food place opened up near your school by someone who thinks that worms are tasty food. Would you like to go there for lunch some day?

1. You buy a Worm Burger for 50¢ and a Worm Cola for 10¢. How much do you spend?

2. Your teacher buys a Worm Burger for 50¢ and Worm Fries for 25¢. How much does she spend in all?

3. Your best friend buys a Worm Cola for 10¢ and Worm Fries for 25¢. How much does he spend?

4. Anne buys a Worm Burger for 50¢. Alice buys Worm Fries for 25¢. Abby buys a Super Worm Burger for 70¢. Your teacher pays for it all. How much does it cost her?

5. Your teacher is hungry. She buys a Worm Burger for 50¢ and a Super Worm Burger for 70¢. How much does she spend?

6. You buy a Worm Cola for 10¢. Your best friend buys a Worm Cola for 10¢. Your teacher buys a Worm Cola, too. How much does it cost for all three Worm Colas?

Skill: Working with Money (subtraction, some regrouping)

Worm Burger Sale

Name ____________________

1. You have 75¢. You buy a Worm Burger. How much money do you have left?

2. A Worm Berry Shake costs 85¢. You have 95¢. How much change do you get?

3. Your teacher bought Worm Fries to go with her lunch. How much change did she get back from 50¢?

4. A Worm Berry Shake costs 85¢. Worm Fries cost 25¢. How much more is a Worm Berry Shake?

5. A Super Worm Burger costs 70¢. You have only 45¢. How much more money do you need?

6. Al bought his lunch at **Worm Burgers.** He spent 95¢. Don spent 40¢ on his lunch. How much more did Al spend?

7. You got 80¢ from your mother to spend at **Worm Burgers.** You spent 60¢. How much do you have left?

Pat's Pet Shop

Skill: Working with Money (addition & subtraction, some regrouping)

Name ____________________

1. You went to Pat's Pet Shop to buy birdseed which cost 59¢ and rabbit food which cost 39¢. What was the sum?

2. You bought a stuffed cat for your pet mouse to play with. It cost 44¢. You gave Pat 50¢. How much change did you get?

3. Your friend bought a bottle of fish food for 39¢ and a can of cat food for 29¢. What was the total?

4. Your teacher bought a bag of seed for her pet bird. It cost 39¢. She gave Pat 75¢. How much change did she get?

5. Your best friend needed a play mouse for his cat. It cost 67¢. He gave Pat 75¢. How much change did he get?

6. Pat sold your teacher a pet mouse for 65¢ and a pet lizard for 34¢. What was the total?

7. The large box of goldfish food costs 98¢. The small size costs 49¢. What is the difference?

Bob Z. Cat

Name ___________________________

Bob Z. Cat is a very odd cat. He is afraid of mice. He loves dogs, and he can't wait to take a bath.

1. Bob Cat has a bottle of bubble bath which cost a quarter and smelly soap which cost a dime. What was the total?

2. Bob Cat has a toy mouse which cost 43¢ and a toy rat which cost 34¢. What was the sum?

3. Bob Cat used to have 86¢. He spent 54¢ on a ball of string. How much does he have left?

4. Bob Cat had 89¢ in his piggy bank. He spent 42¢. How much was left?

5. Bob Cat was given catnip which cost 56¢ and a toy mouse which cost 45¢. What was the sum?

6. Bob Z. Cat can have the large can of cat food which costs 69¢ or the small can which costs 51¢. What is the difference?

Animal Trivia

Skill: Metric Measurement (centimeters, some regrouping)

Name ____________________

1. The house mouse has a tail 10 cm long. The pack rat has a tail 19 cm long. How much longer is the rat's tail?

2. A chipmunk is 24 cm long. A gopher is 19 cm long. How much longer is the chipmunk?

3. The gray squirrel has a tail 24 cm long. The red squirrel has a tail 16 cm long. How much longer is the tail of the gray squirrel?

4. The spotted bat has a body that is 11 cm long. His tail is 5 cm long. How long is the bat from the tip of his nose to the tip of his tail?

5. The jackrabbit has a back foot that is 17 cm long. The back foot of the cottontail rabbit is 10 cm long. How much longer is the back foot of the jackrabbit?

6. The back foot of a polar bear is 33 cm long. The back foot of a black bear is 37 cm long. How much longer is the black bear's foot?

7. The body of a pocket mouse is 20 cm long. His tail is 14 cm long. How long is the pocket mouse from the tip of his nose to the tip of his tail?

Lizzy the Lizard

Name ____________________

1. Lizzy the Lizard ate 3 beetles for breakfast. She ate 3 times as many for lunch. How many did she eat for lunch?

2. Lizzy ate 4 plates of beetles. Each plate had 3 beetles. How many beetles did she eat?

3. Lizzy the Lizard caught 5 bugs. Her friend Dizzy caught 3 times as many bugs. How many bugs did Dizzy catch?

4. Lizzy had 3 piles of bugs. Each pile had 4 bugs. How many bugs did she have?

5. Lizzy had 2 crickets for a snack. She had 4 snacks in a day. How many crickets did she eat?

6. Lizzy caught 4 ants. Her friend Dizzy caught 2 times as many ants. How many ants did Dizzy catch?

7. Lizzy counted 6 beetles in an hour. She counted 2 times as many in the next hour. How many did she count in the next hour?

Super Sale at Pat's Pet Place

Name ______________________________

1. Pat had 30 mice when the sale started. He sold 10 mice in the morning and 9 mice in the afternoon. How many mice were left?

2. Pat had 50 goldfish when the sale started. He sold 12 goldfish to Sam and 14 goldfish to Don. How many fish does Pat have now?

3. Pat started the day with 40 crickets. He sold 20 crickets, and 12 more crickets got away. How many does Pat have now?

4. Pat sold a guppy to Larry for 23¢ and a goldfish for 34¢. Larry gave Pat 3 quarters. How much change should Pat give Larry?

5. Pat had 18 chicks when the sale began. He sold 13 chicks to one person and 5 chicks to another. How many chicks are left?

6. Pat sold 8 rabbits to Fred and 9 rabbits to Fanny. He started out with 20 rabbits. How many rabbits are still there?

7. Pat sold a can of cat food for 23¢ and a can of mouse food for 45¢ to John. John gave Pat 95¢. How much change did John get?

Mean Monster's Diet

Name ______________________________

Mean Monster has to go on a diet. He is so fat he popped all the buttons off his shirt. Help him choose the right piece of food.

1. Mean Monster may have 1/4 of this chocolate pie. Color in 1/4 of the pie.

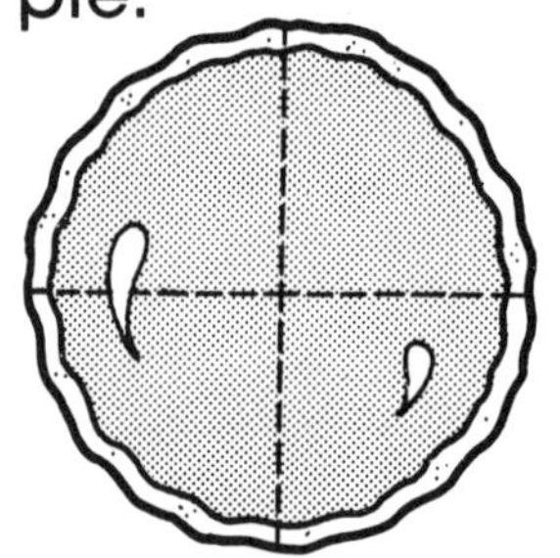

2. Mean Monster may eat 1/3 of this pizza. Color in 1/3 of the pizza.

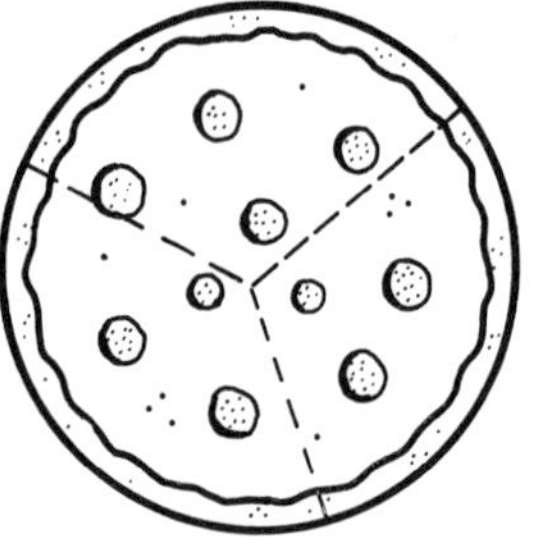

3. For a snack, he wants 1/3 of this chocolate cake. Color in 1/3 of the cake.

4. For lunch, Mean Monster gets 1/2 of the sandwich. Color in 1/2 of the sandwich.

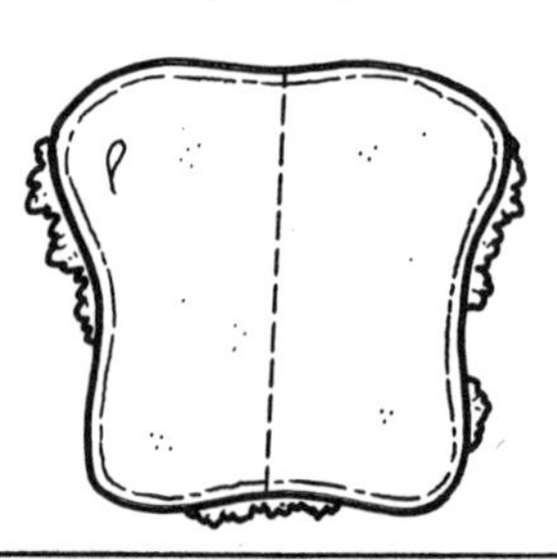

5. For an evening snack, he can have 1/4 of the candy bar. Color in 1/4 of the candy bar.

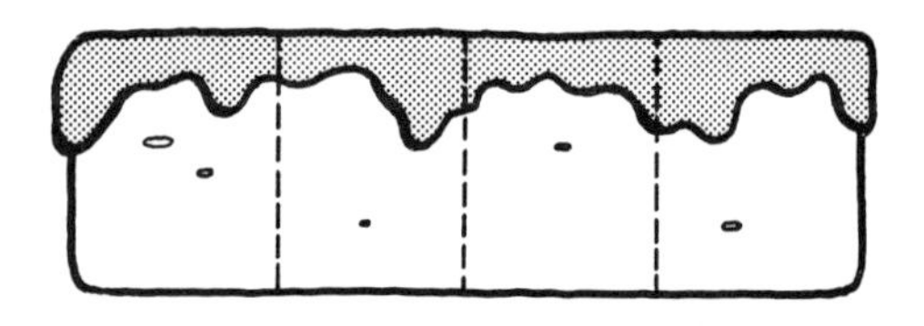

6. He ate 1/2 of the apple for lunch. Color in 1/2 of the apple.

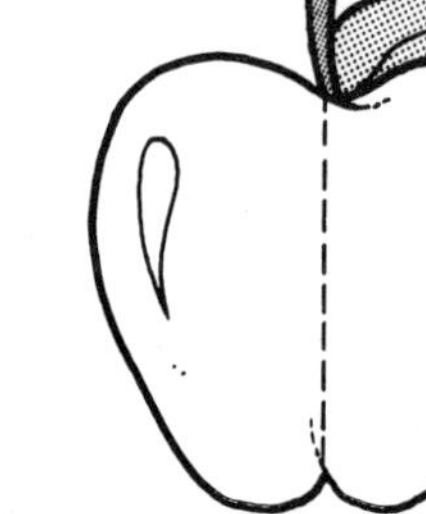

Answer Key

Skill: Counting money - penny

Henny Penny

Name ____________________

1¢ 1 penny 1 cent

How much money?

Example

1¢ 1¢ 1¢ 1¢ 1¢ = 5 ¢

1¢ 1¢ 1¢ 1¢ 1¢ 1¢ = 6 ¢

1¢ 1¢ = 2 ¢

1¢ 1¢ 1¢ = 3 ¢

1¢ 1¢ 1¢ 1¢ = 4 ¢

1¢ 1¢ 1¢ 1¢ 1¢ 1¢ 1¢ 1¢ = 8 ¢

1¢ 1¢ 1¢ 1¢ 1¢ 1¢ 1¢ = 7 ¢

Page 1

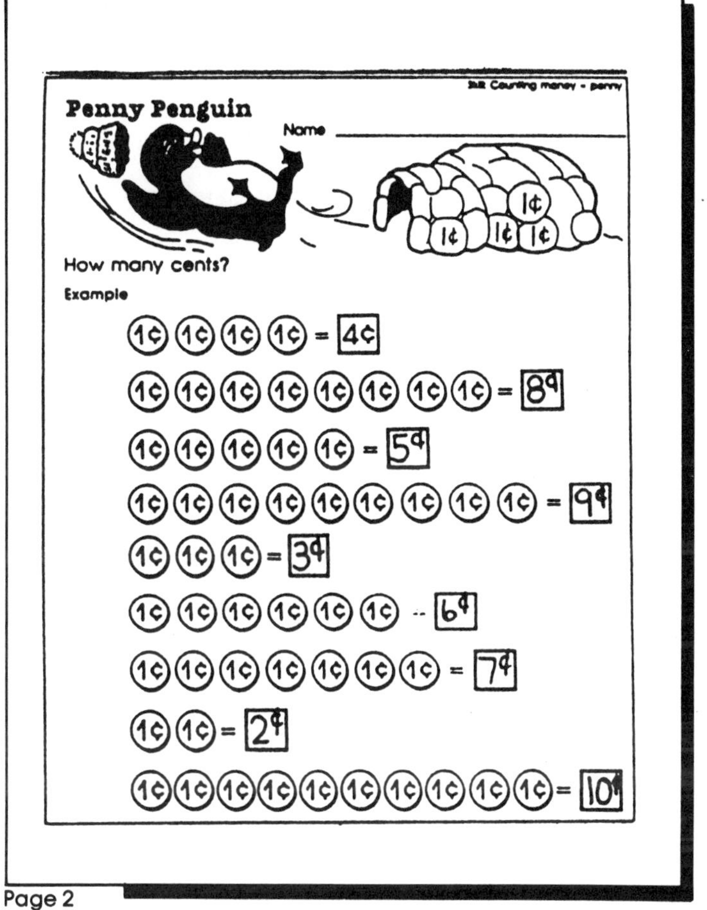

Skill: Counting money - penny

Penny Penguin

Name ____________________

How many cents?

Example

1¢ 1¢ 1¢ 1¢ = 4¢

1¢ 1¢ 1¢ 1¢ 1¢ 1¢ 1¢ 1¢ = 8¢

1¢ 1¢ 1¢ 1¢ 1¢ = 5¢

1¢ 1¢ 1¢ 1¢ 1¢ 1¢ 1¢ 1¢ 1¢ = 9¢

1¢ 1¢ 1¢ = 3¢

1¢ 1¢ 1¢ 1¢ 1¢ 1¢ = 6¢

1¢ 1¢ 1¢ 1¢ 1¢ 1¢ 1¢ = 7¢

1¢ 1¢ = 2¢

1¢ 1¢ 1¢ 1¢ 1¢ 1¢ 1¢ 1¢ 1¢ 1¢ = 10¢

Page 2

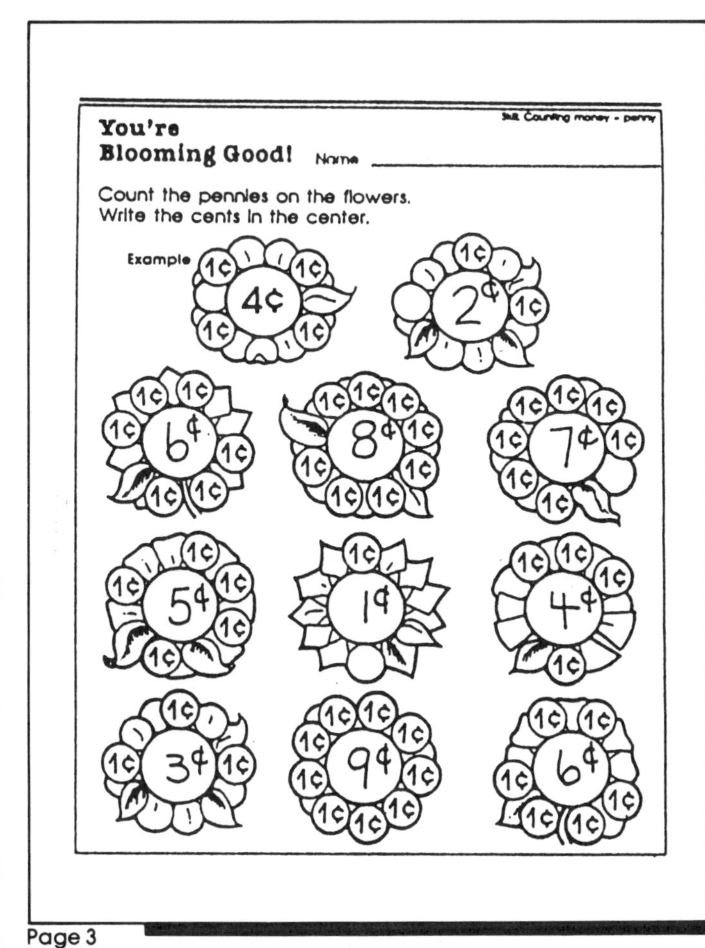

Skill: Counting money - penny

You're Blooming Good!

Name ____________________

Count the pennies on the flowers.
Write the cents in the center.

Example

4¢ 2¢

6¢ 8¢ 7¢

5¢ 1¢ 4¢

3¢ 9¢ 6¢

Page 3

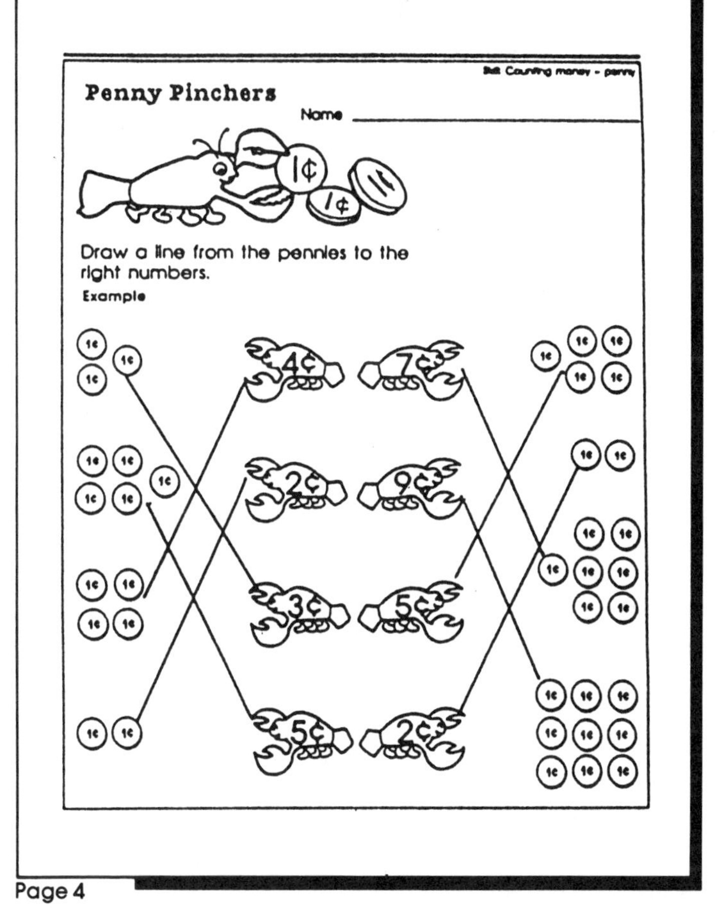

Skill: Counting money - penny

Penny Pinchers

Name ____________________

Draw a line from the pennies to the right numbers.

Example

4¢ 7¢

2¢ 9¢

3¢ 5¢

5¢ 2¢

Page 4

Answer Key

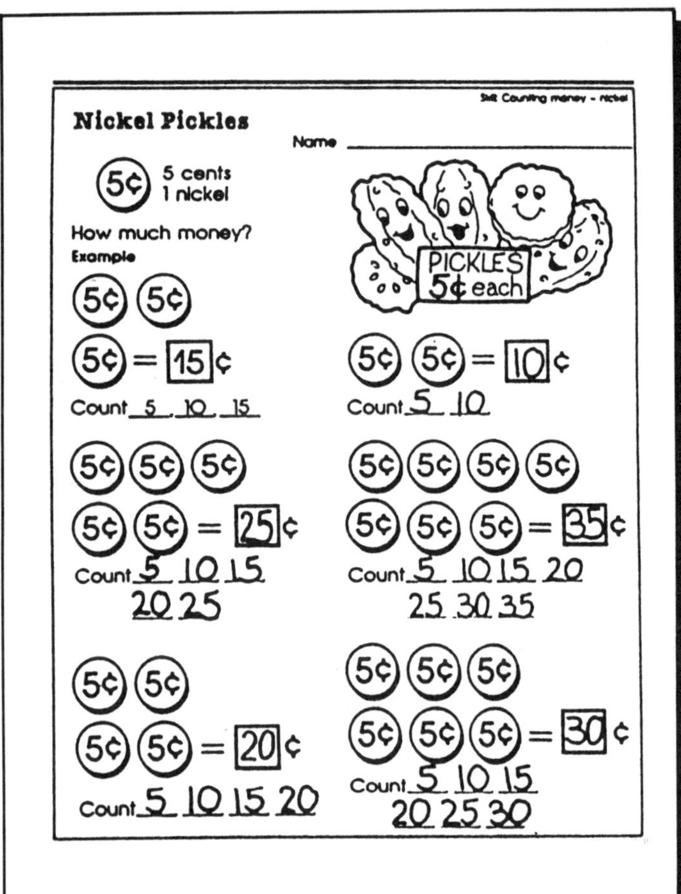

Page 5

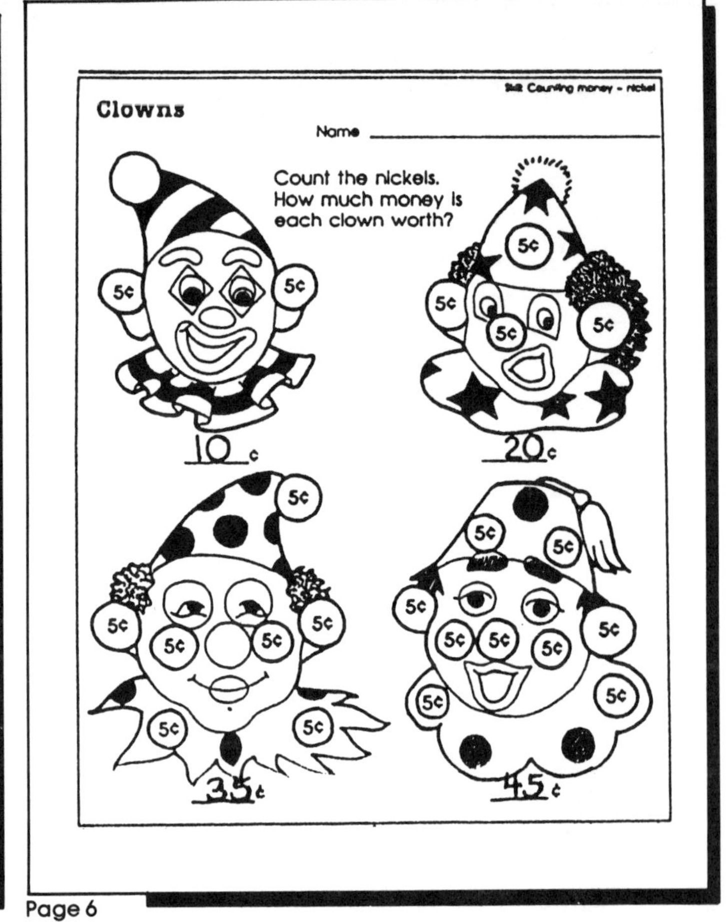

Page 6

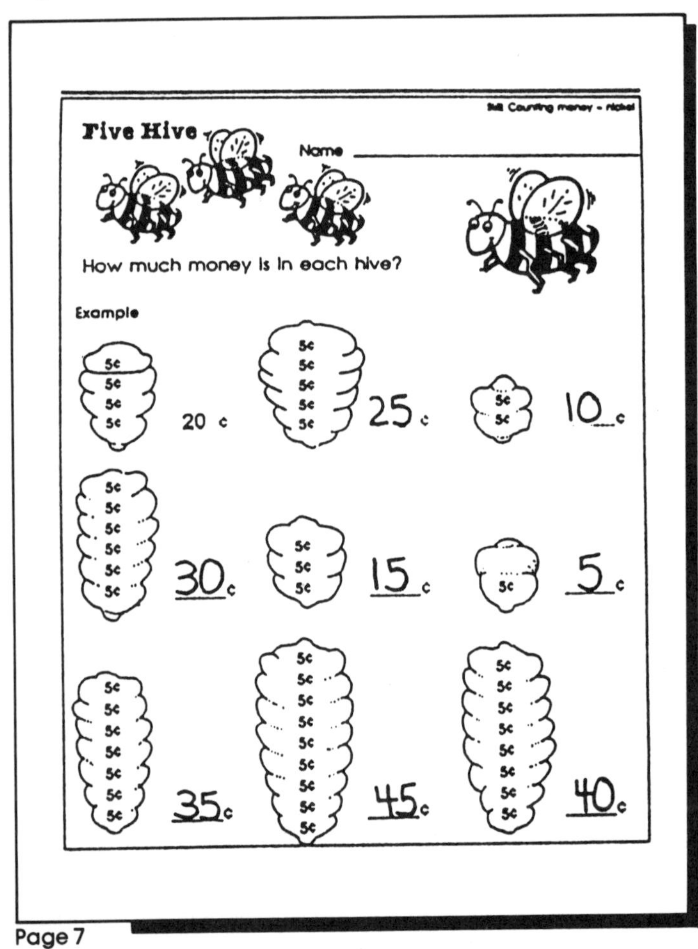

Page 7

Page 8

Answer Key

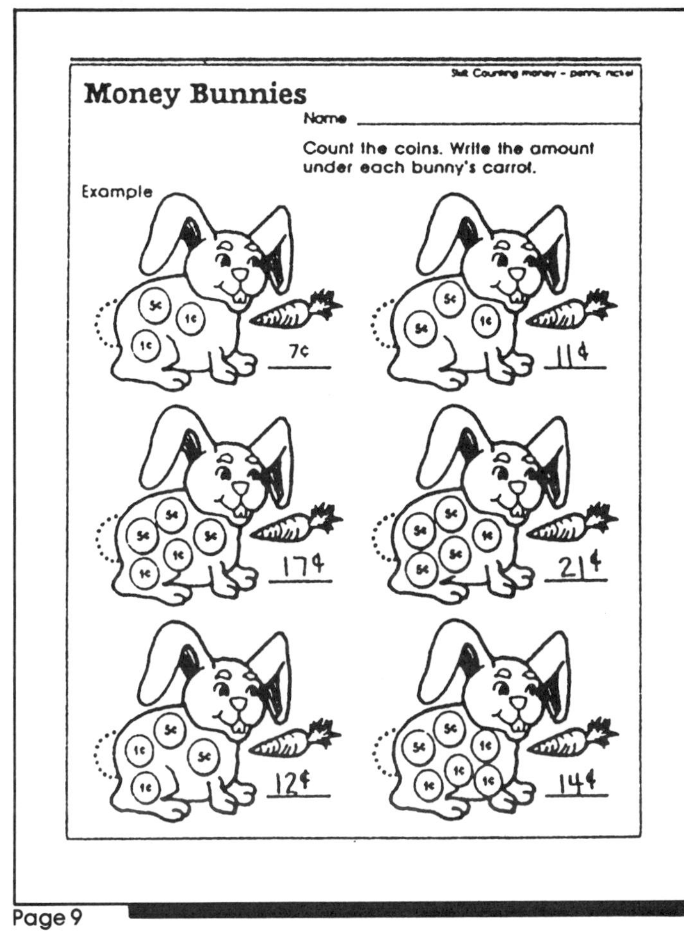

Page 9

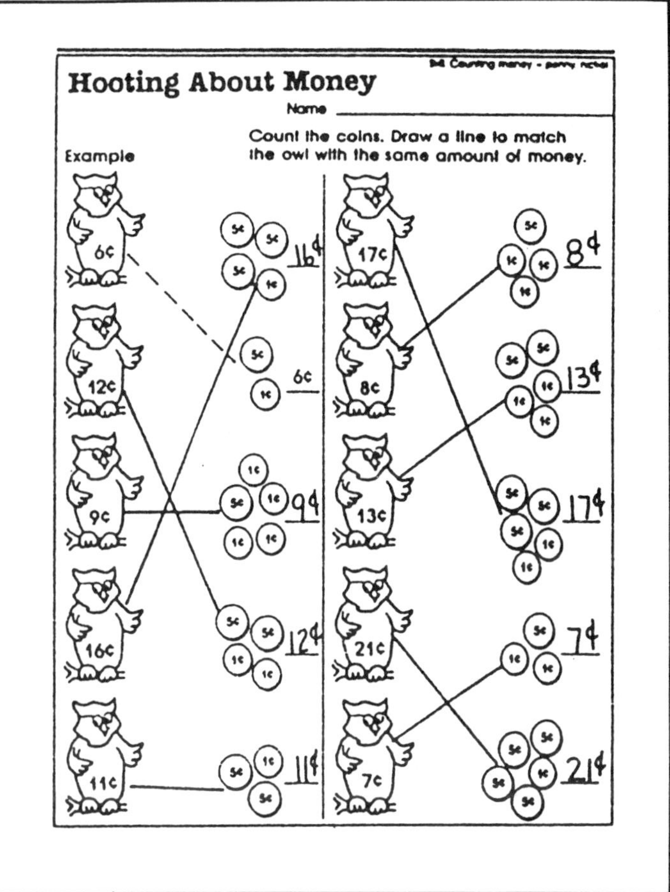

Page 10

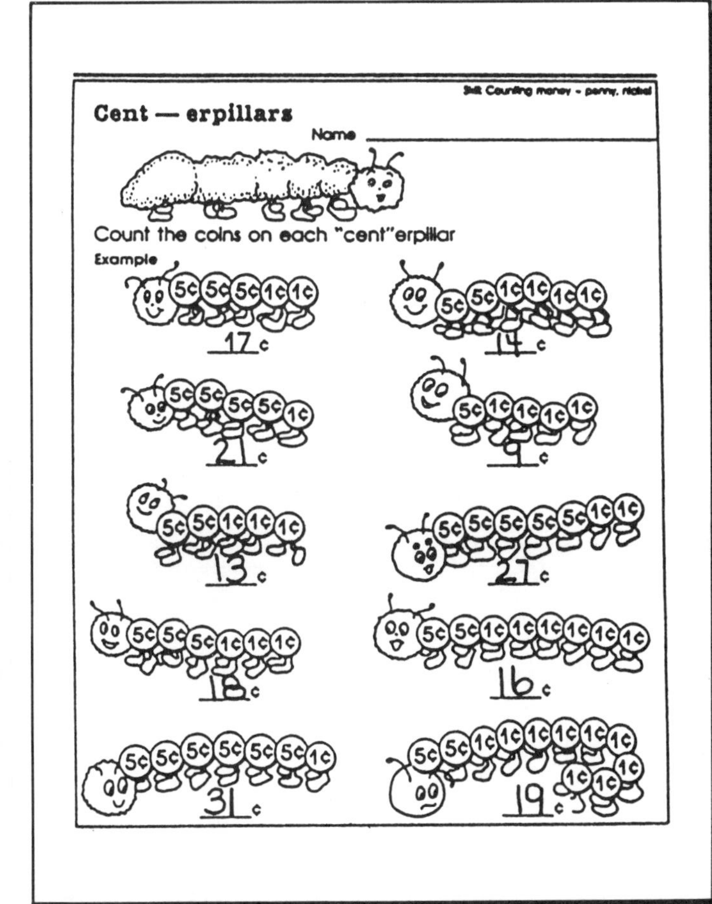

Page 11

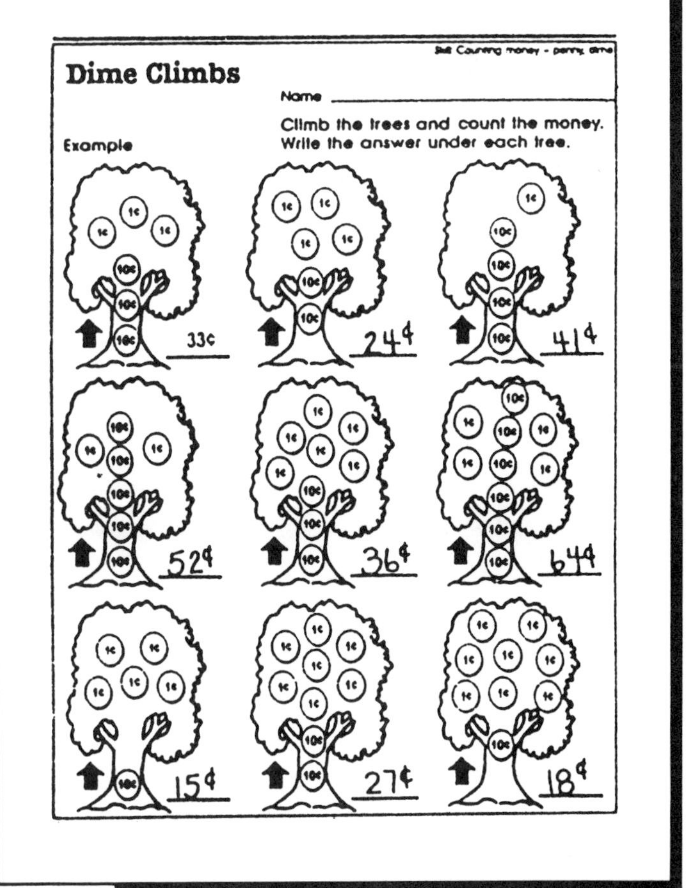

Page 12

Answer Key

Page 13

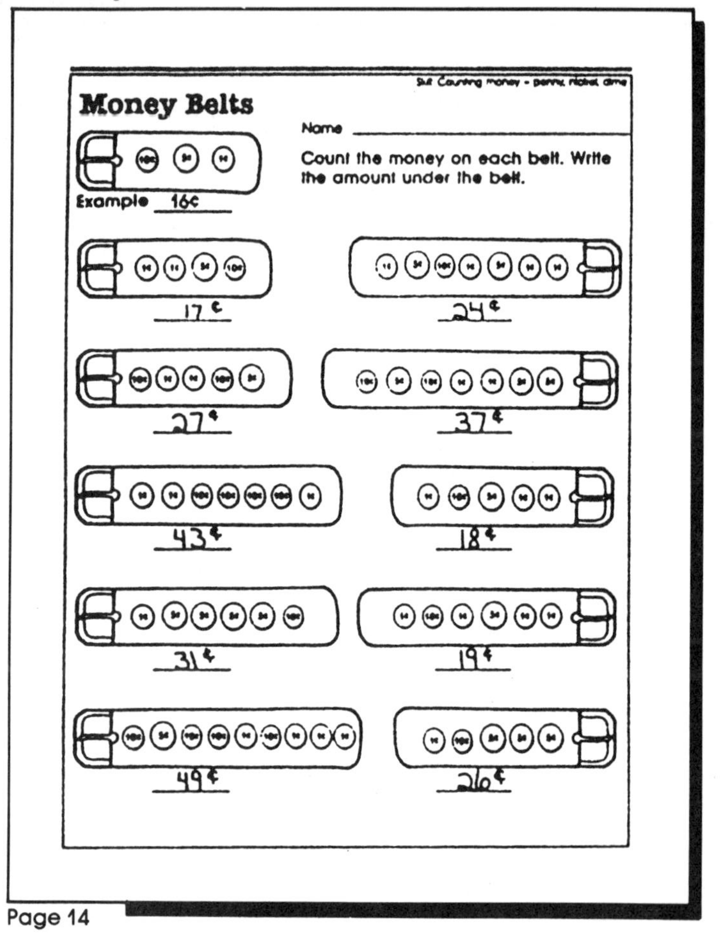

Page 14

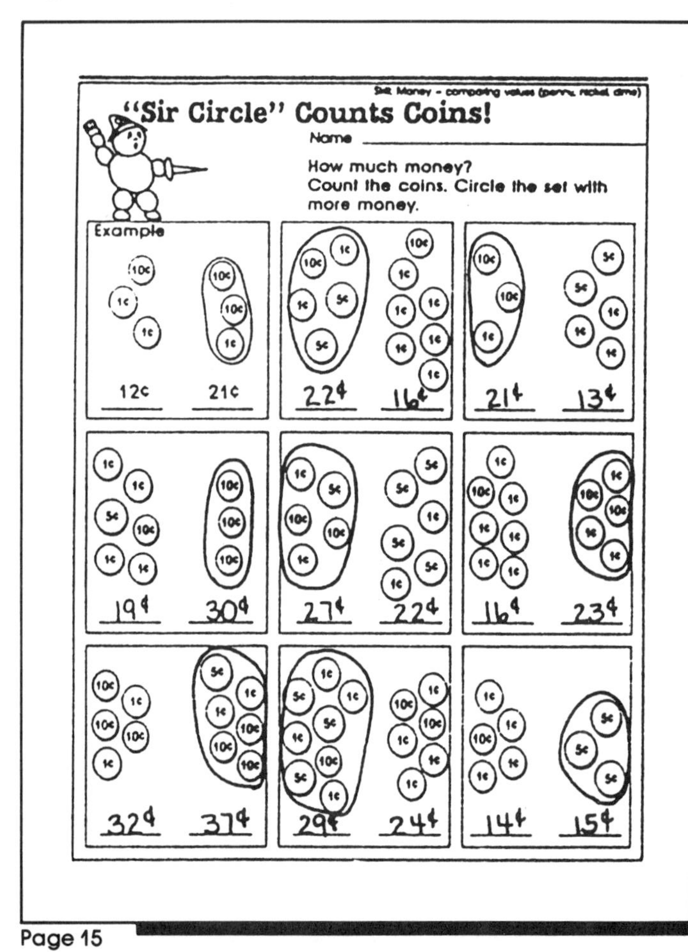

Page 15

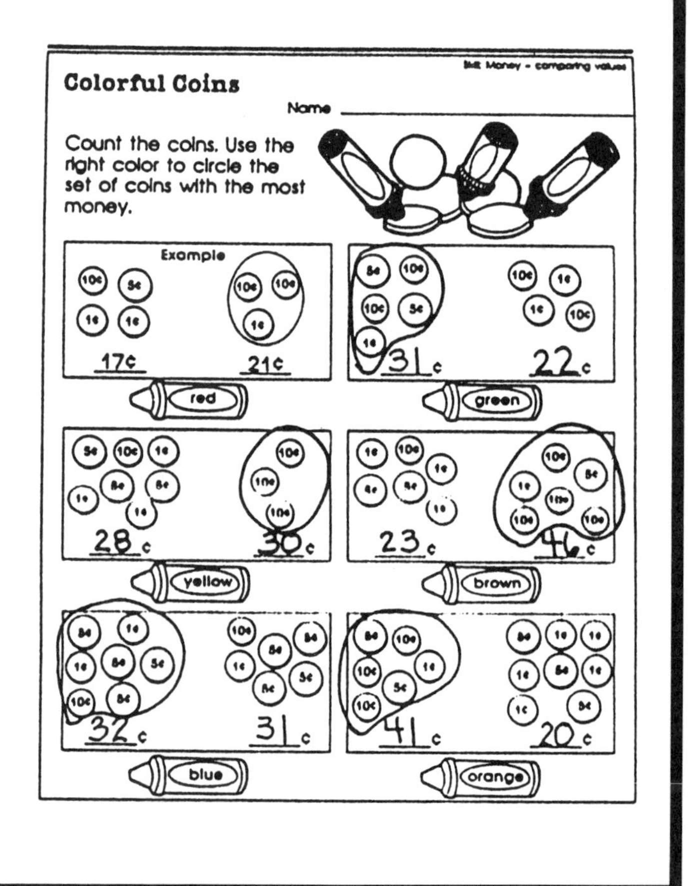

Page 16

Answer Key

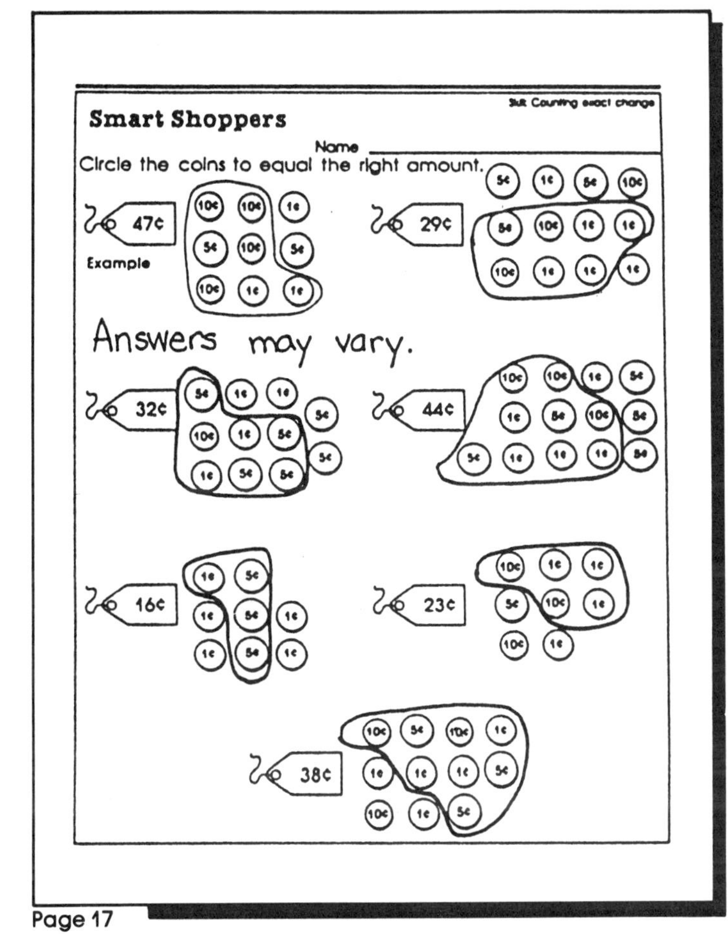

Smart Shoppers

Skill: Counting exact change

Name ______________

Circle the coins to equal the right amount.

Example: 47¢ — 29¢

Answers may vary.

32¢ — 44¢

16¢ — 23¢

38¢

Page 17

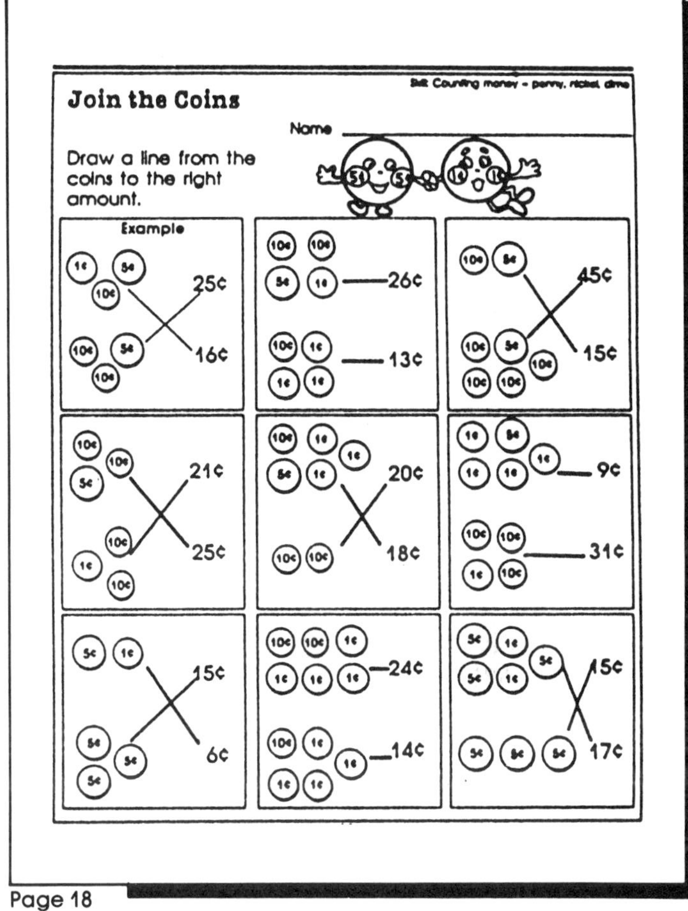

Join the Coins

Skill: Counting money - penny, nickel, dime

Name ______________

Draw a line from the coins to the right amount.

Example: 25¢, 16¢ — 26¢, 13¢ — 45¢, 15¢

21¢, 25¢ — 20¢, 18¢ — 9¢, 31¢

15¢, 6¢ — 24¢, 14¢ — 15¢, 17¢

Page 18

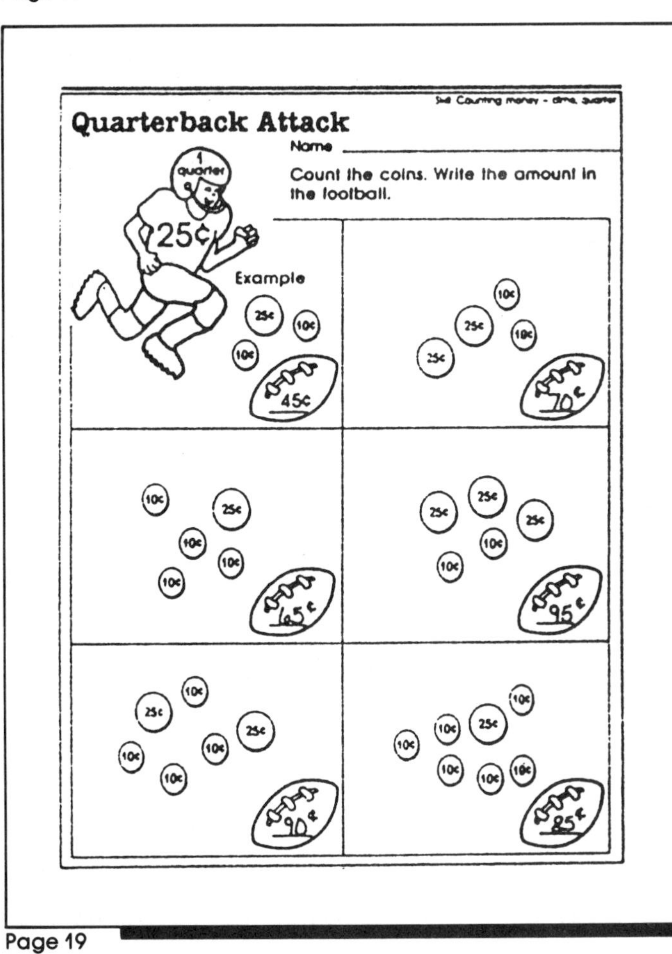

Quarterback Attack

Skill: Counting money - dime, quarter

Name ______________

Count the coins. Write the amount in the football.

1 quarter 25¢

Example: 45¢

70¢

65¢ — 95¢

90¢ — 85¢

Page 19

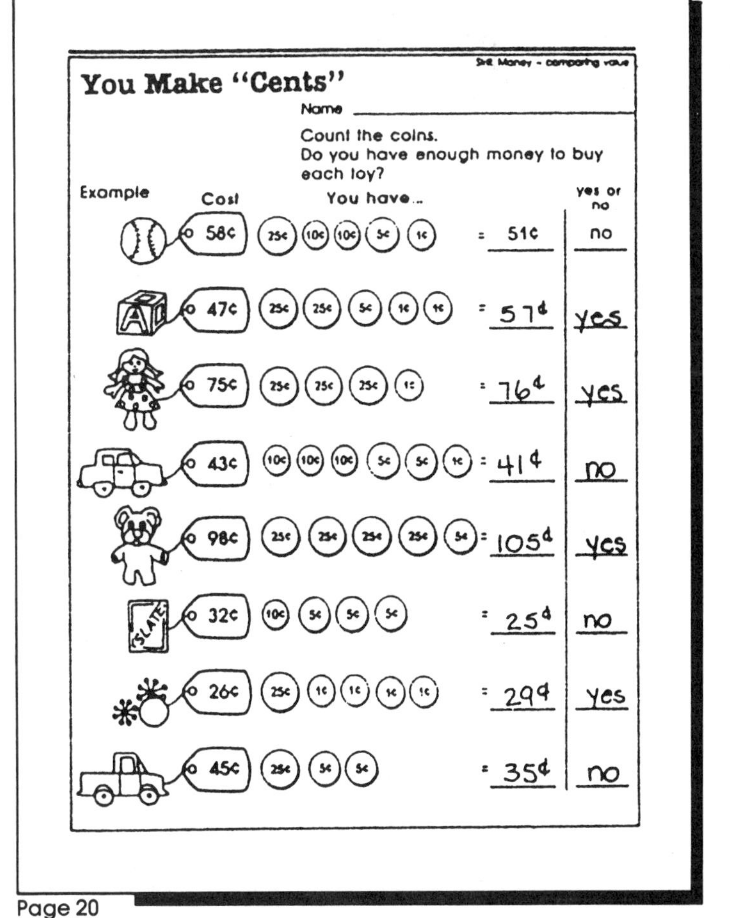

You Make "Cents"

Skill: Money - comparing value

Name ______________

Count the coins.
Do you have enough money to buy each toy?

Example	Cost	You have...		yes or no
	58¢	25¢ 10¢ 10¢ 5¢ 1¢	= 51¢	no
	47¢	25¢ 25¢ 5¢ 1¢ 1¢	= 57¢	yes
	75¢	25¢ 25¢ 25¢ 1¢	= 76¢	yes
	43¢	10¢ 10¢ 10¢ 5¢ 5¢ 1¢	= 41¢	no
	98¢	25¢ 25¢ 25¢ 25¢ 5¢	= 105¢	yes
	32¢	10¢ 5¢ 5¢ 5¢	= 25¢	no
	26¢	25¢ 1¢ 1¢ 1¢ 1¢	= 29¢	yes
	45¢	25¢ 5¢ 5¢	= 35¢	no

Page 20

Answer Key

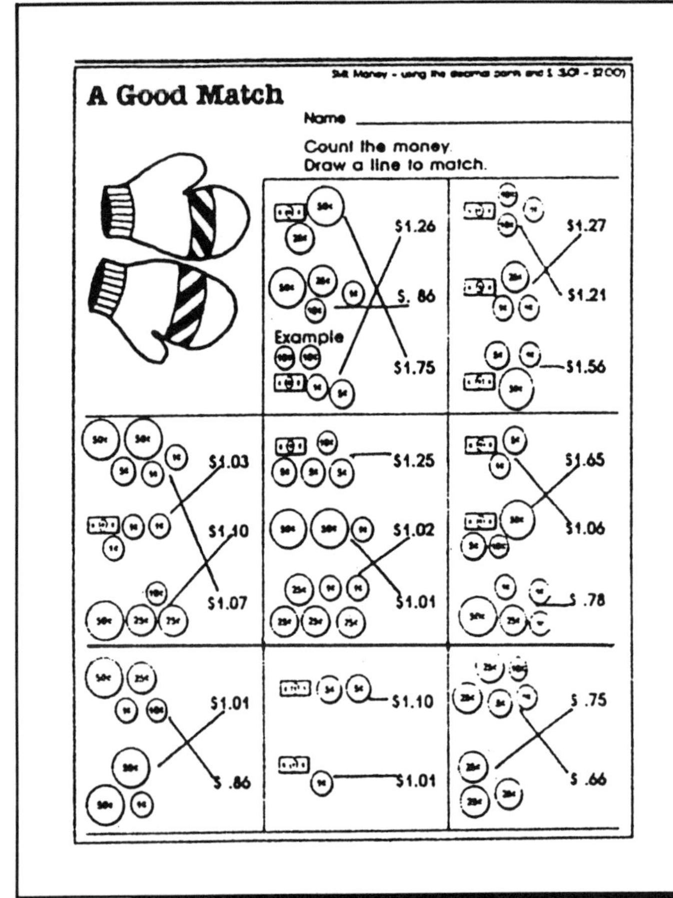

A Good Match

Skill: Money - using the decimal point and $ .50 - $2.00

Name

Count the money.
Draw a line to match.

Example

$1.26 · $.86 · $1.75 · $1.27 · $1.21 · $1.56

$1.03 · $1.10 · $1.07 · $1.25 · $1.02 · $1.01 · $1.65 · $1.06 · $.78

$1.01 · $.86 · $1.10 · $1.01 · $.75 · $.66

Page 21

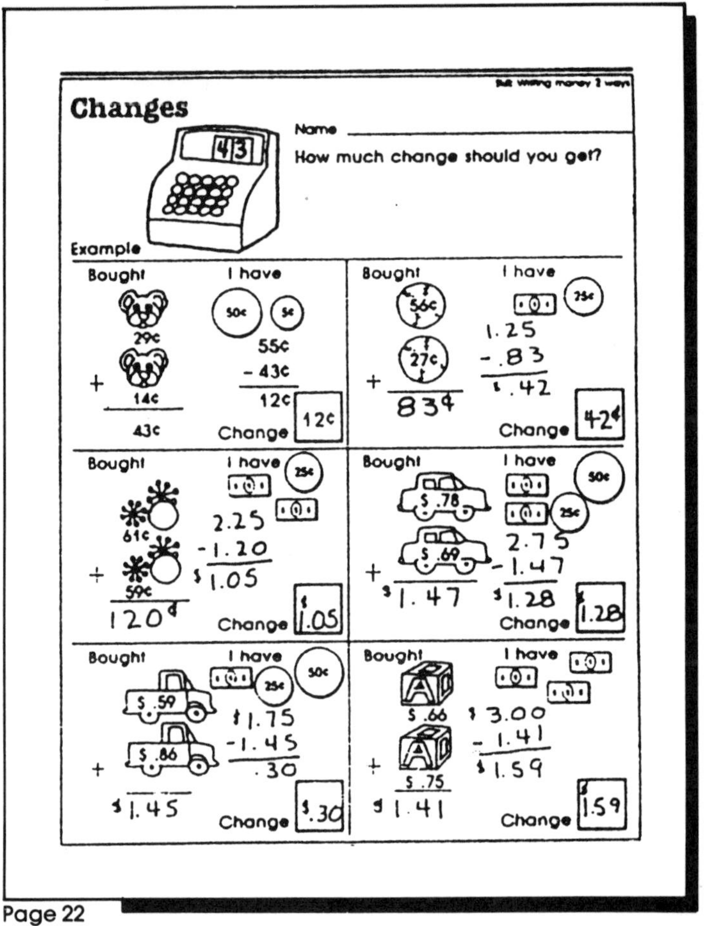

Changes

Skill: Writing money 2 ways

Name

How much change should you get?

Example

Bought	I have	Change
29¢ + 14¢ = 43¢	55¢ − 43¢ = 12¢	12¢
56¢ + 27¢ = 83¢	1.25 − .83 = $.42	42¢
61¢ + 59¢ = 120¢	2.25 − 1.20 = $1.05	$1.05
$.78 + $.69 = $1.47	2.75 − 1.47 = $1.28	$1.28
$.59 + $.86 = $1.45	$1.75 − 1.45 = .30	$.30
$.66 + $.75 = $1.41	$3.00 − 1.41 = $1.59	$1.59

Page 22

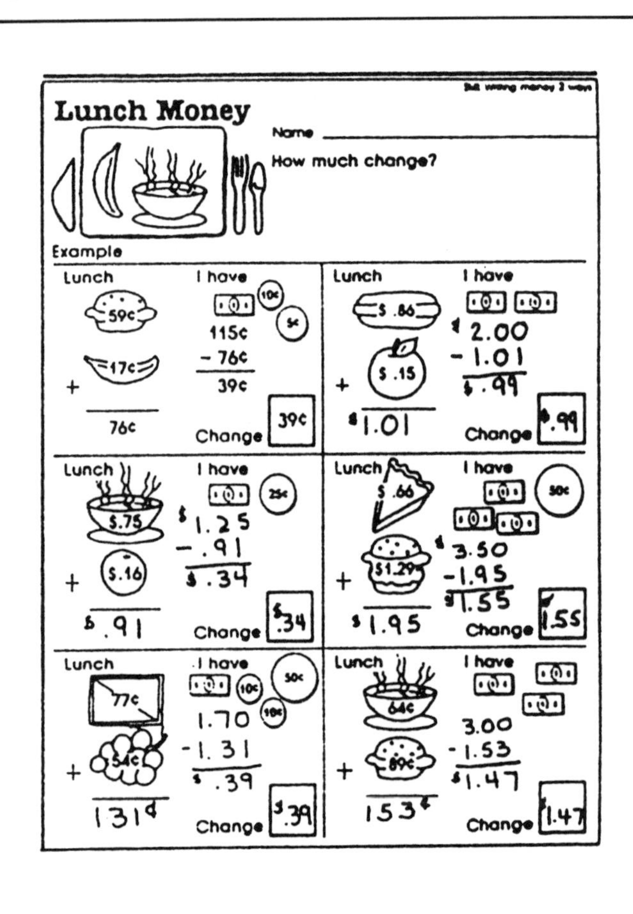

Lunch Money

Skill: Writing money 2 ways

Name

How much change?

Example

Lunch	I have	Change
59¢ + 17¢ = 76¢	115¢ − 76¢ = 39¢	39¢
$.86 + $.15 = $1.01	$2.00 − 1.01 = $.99	$.99
$.75 + $.16 = $.91	$1.25 − .91 = $.34	$.34
$.66 + $1.29 = $1.95	$3.50 − 1.95 = $1.55	$1.55
77¢ + 54¢ = 131¢	1.70 − 1.31 = $.39	$.39
64¢ + 89¢ = 153¢	3.00 − 1.53 = $1.47	$1.47

Page 23

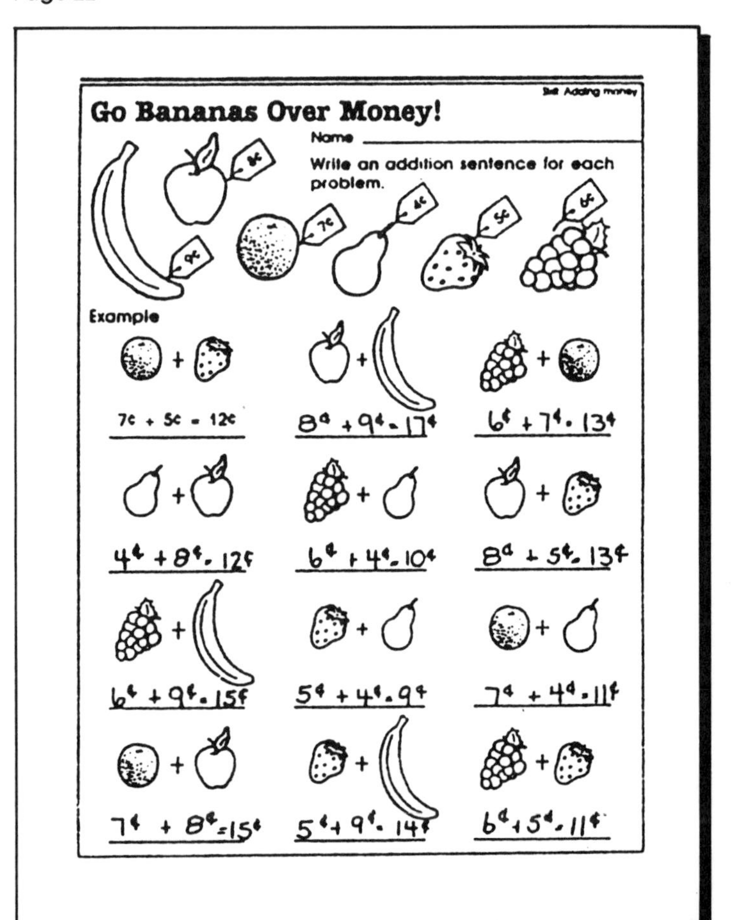

Go Bananas Over Money!

Skill: Adding money

Name

Write an addition sentence for each problem.

Example

7¢ + 5¢ = 12¢ · 8¢ + 9¢ = 17¢ · 6¢ + 7¢ = 13¢

4¢ + 8¢ = 12¢ · 6¢ + 4¢ = 10¢ · 8¢ + 5¢ = 13¢

6¢ + 9¢ = 15¢ · 5¢ + 4¢ = 9¢ · 7¢ + 4¢ = 11¢

7¢ + 8¢ = 15¢ · 5¢ + 9¢ = 14¢ · 6¢ + 5¢ = 11¢

Page 24

Answer Key

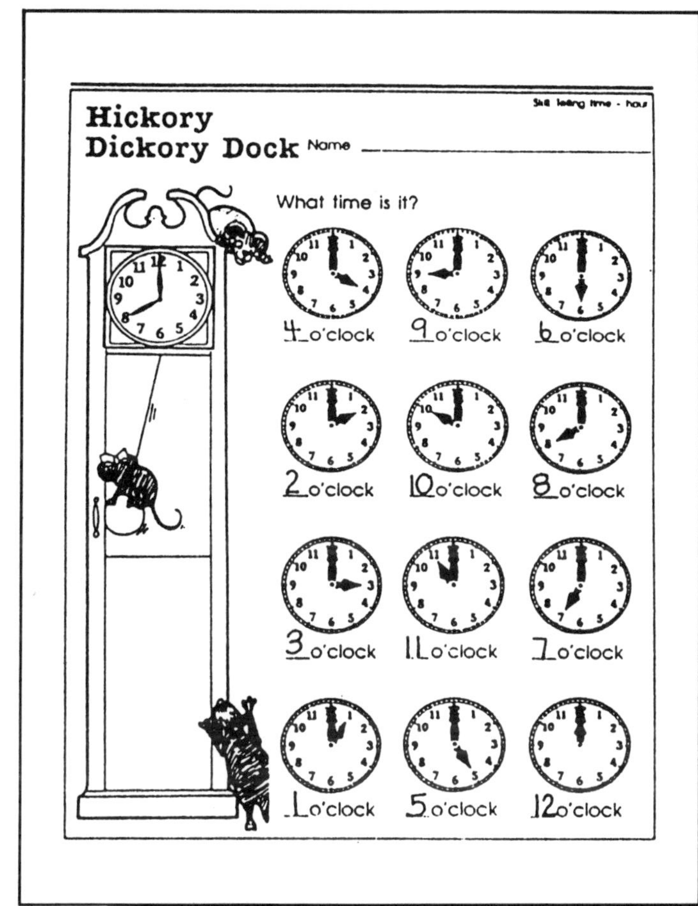

Page 25

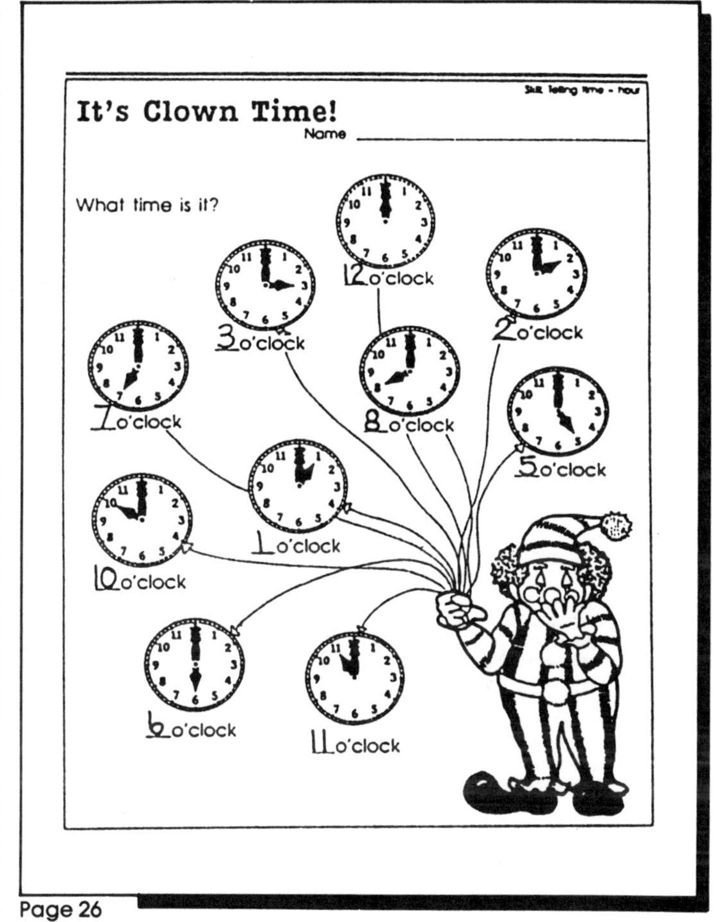

Page 26

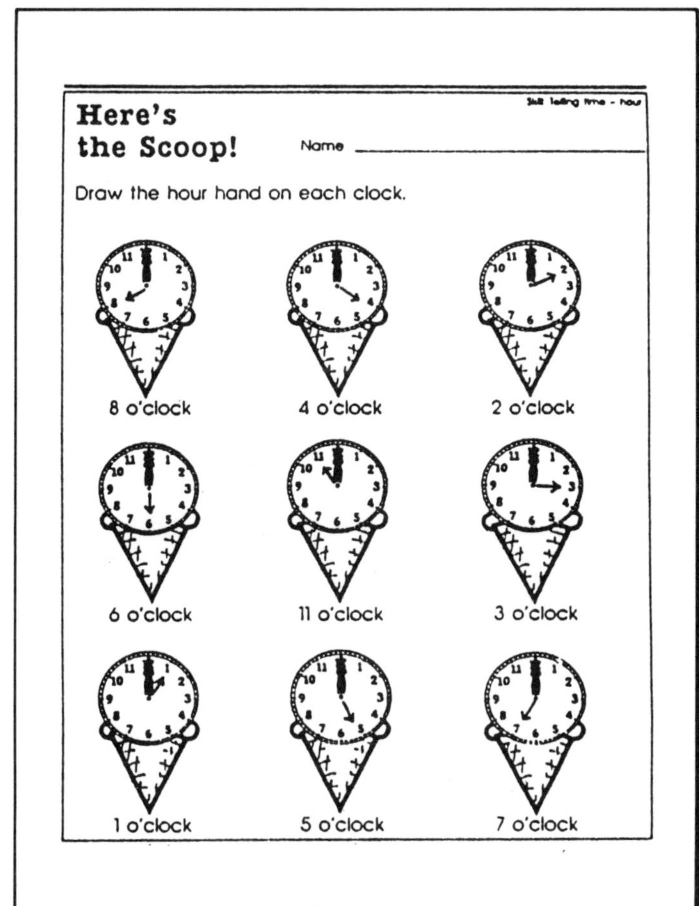

Page 27

Page 28

Answer Key

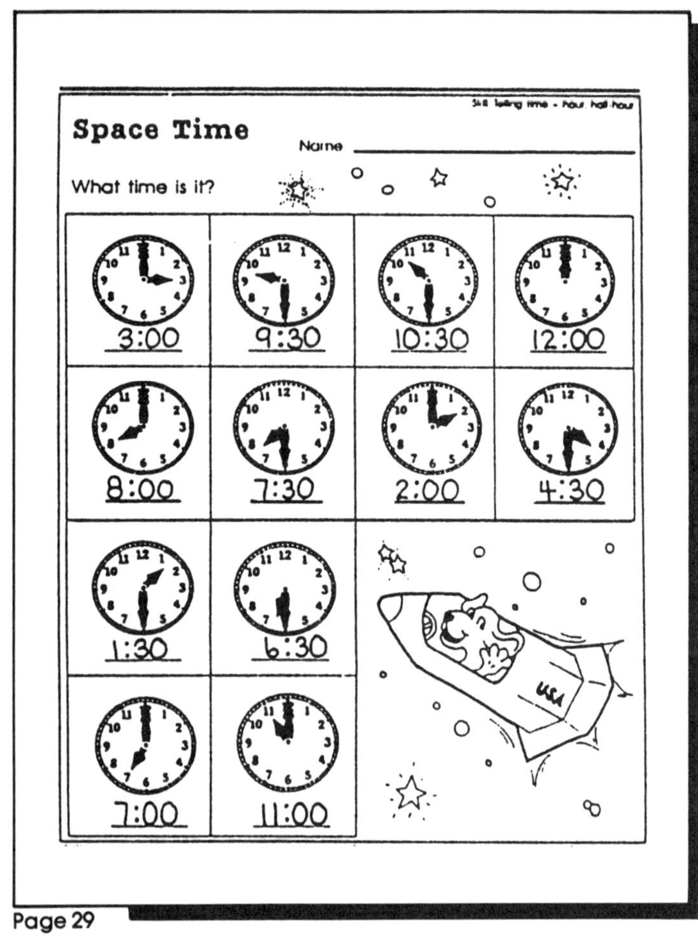

Page 29

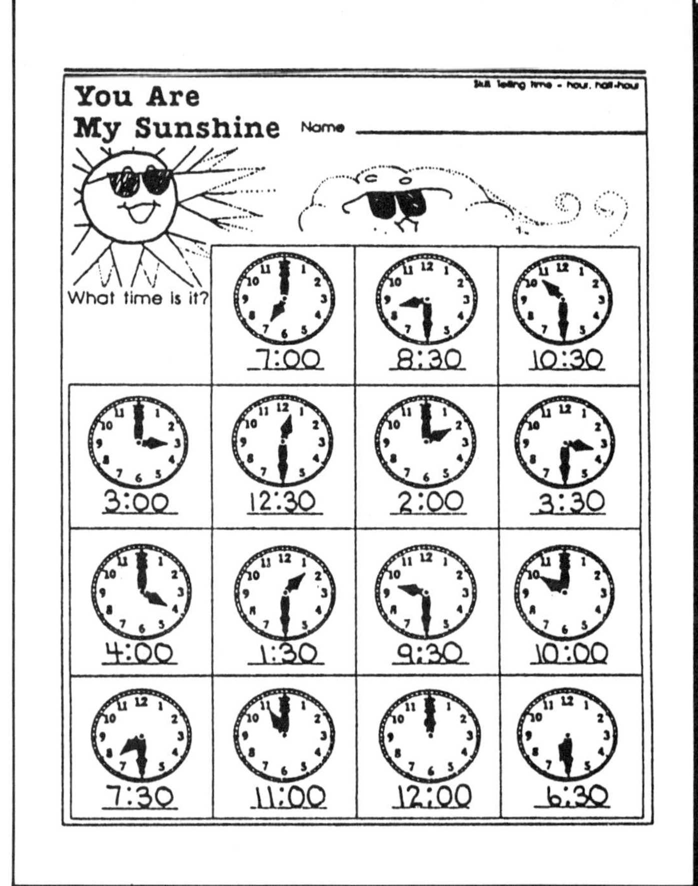

Page 30

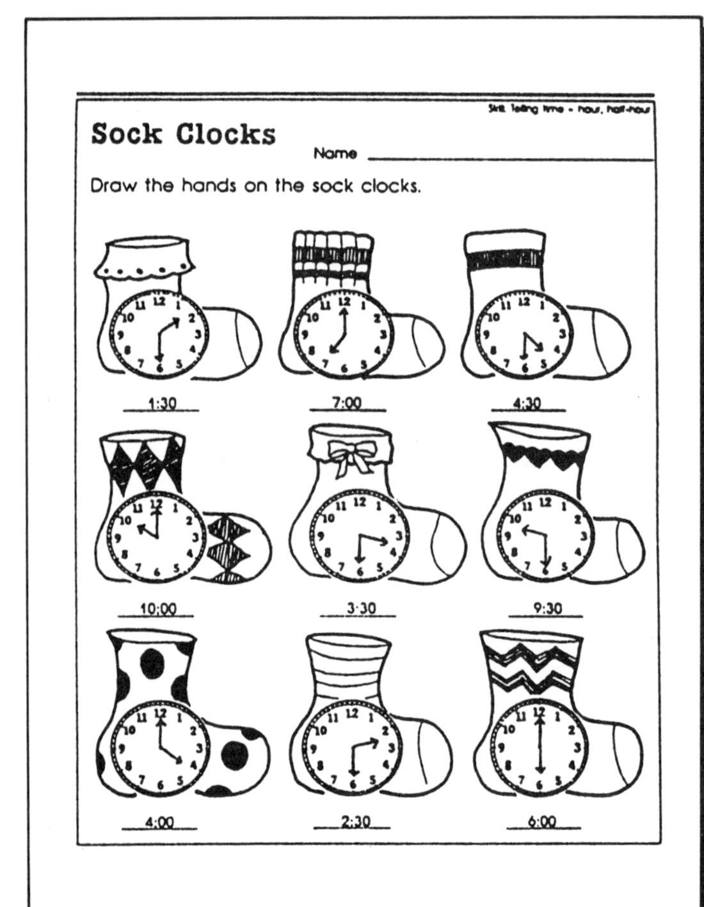

Page 31

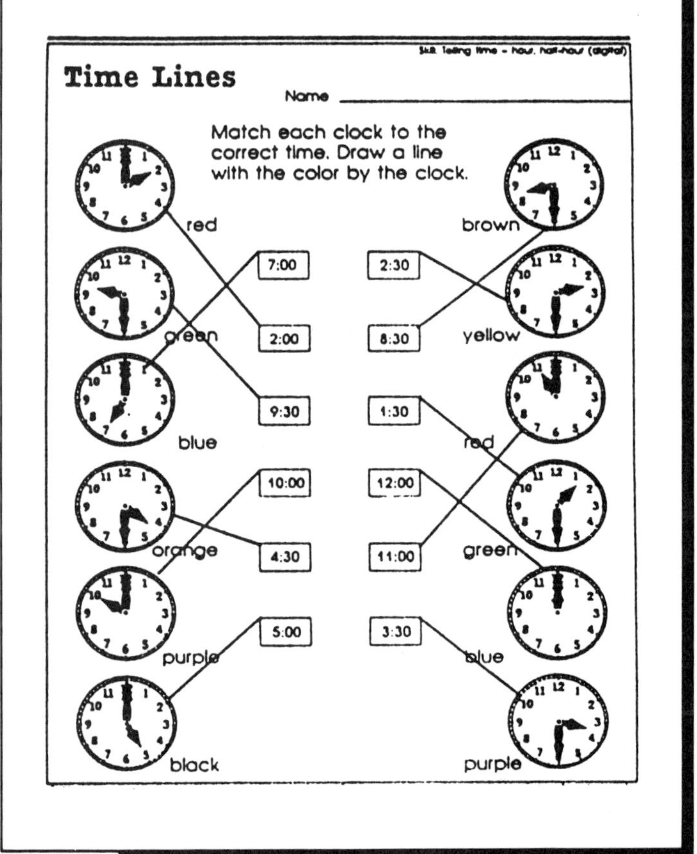

Page 32

Answer Key

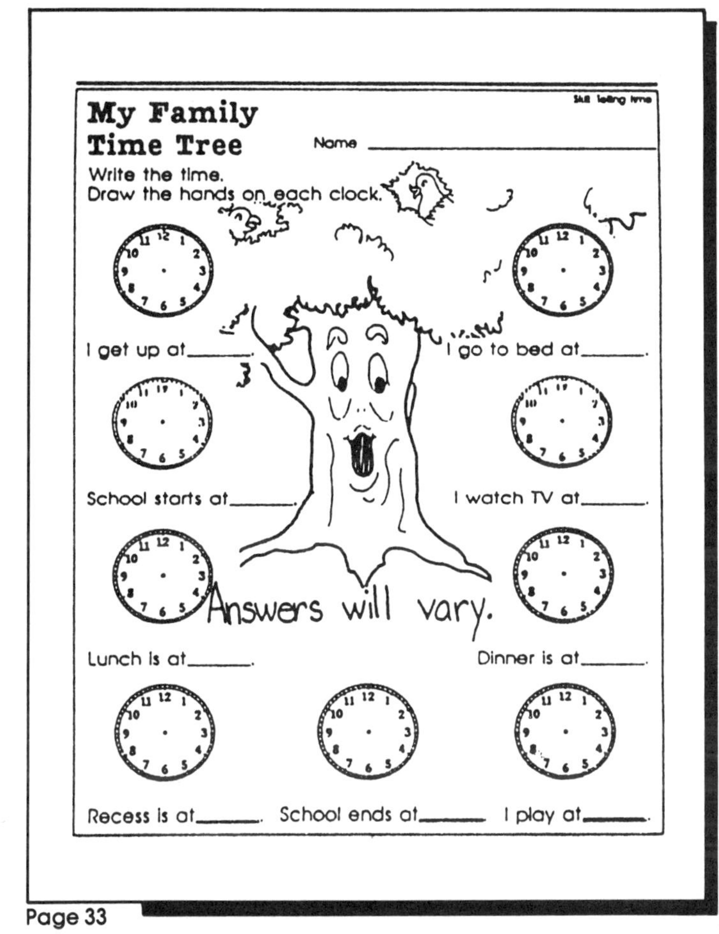

Page 33

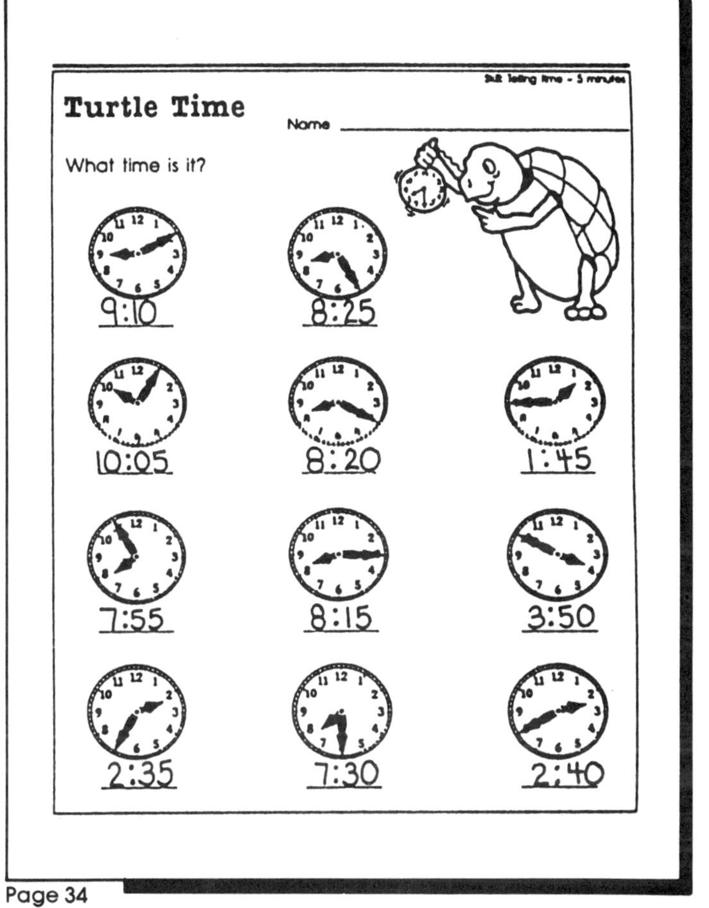

Page 34

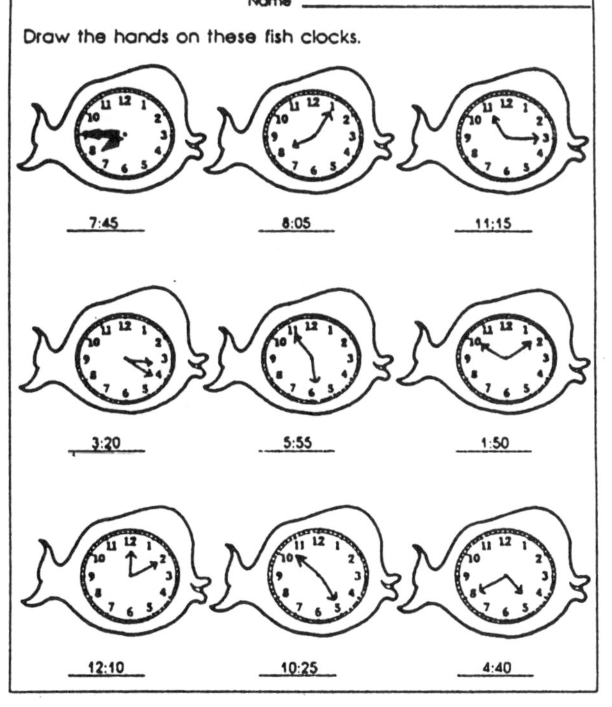

Page 35

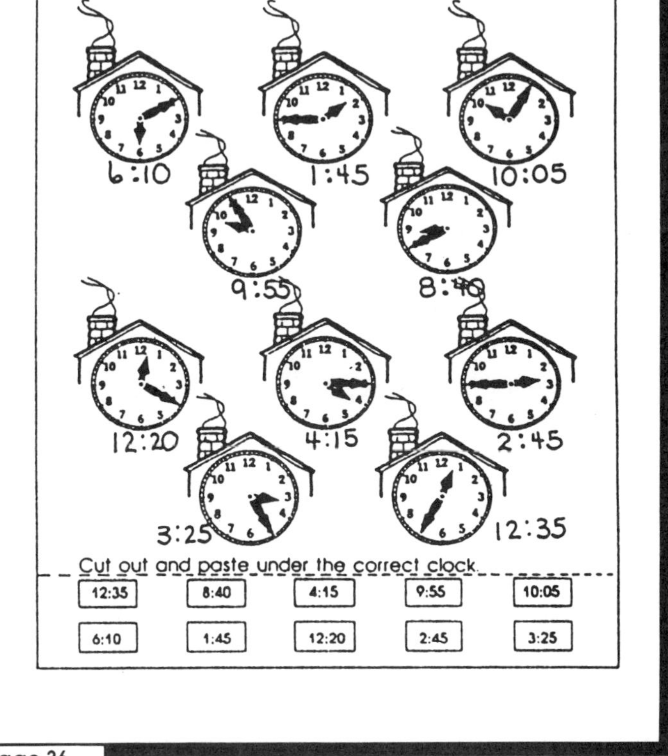

Page 36

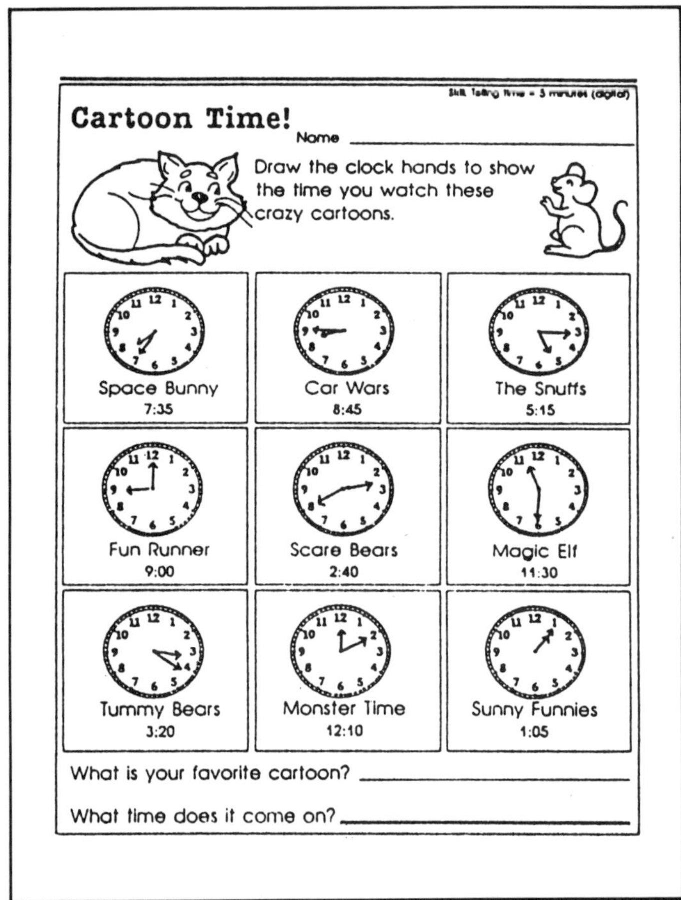

Page 37

Page 38

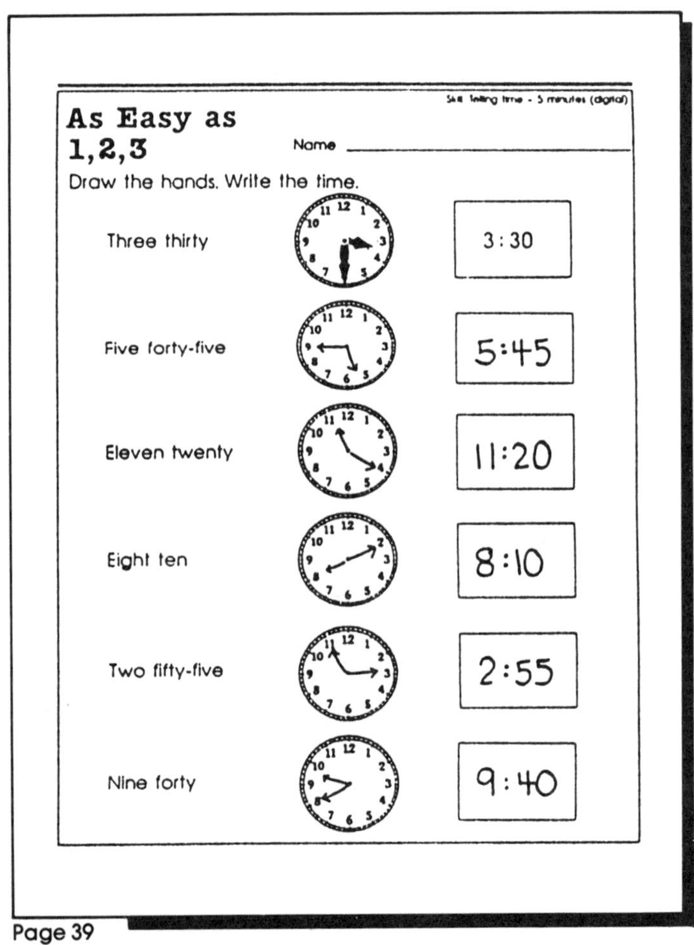

Page 39

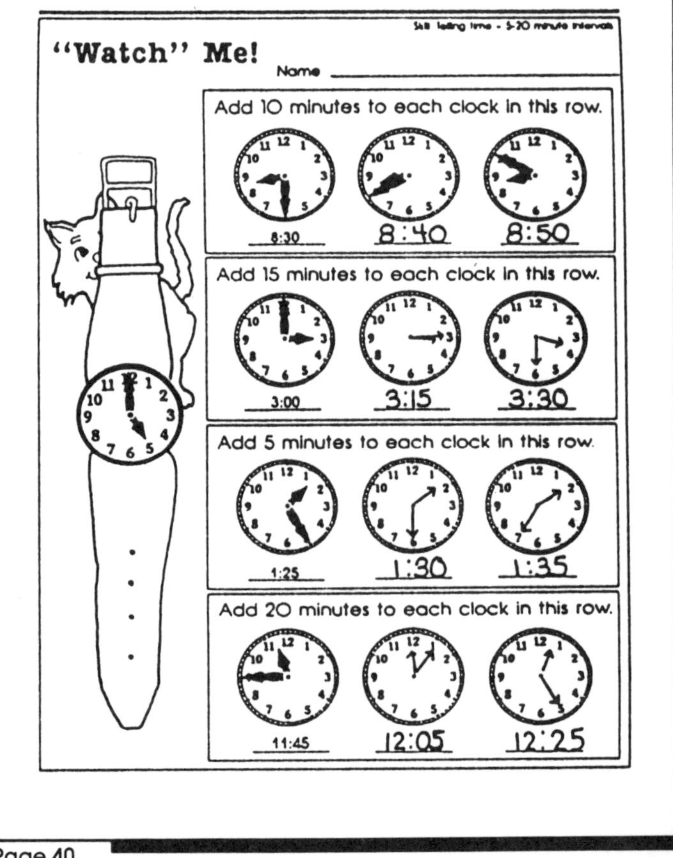

Page 40

Answer Key

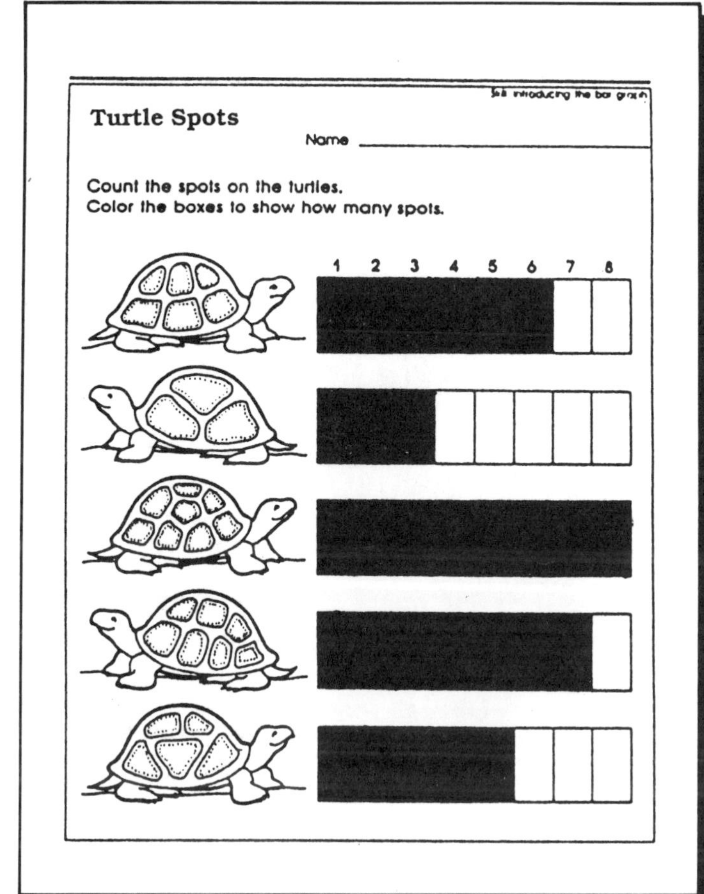

Page 41

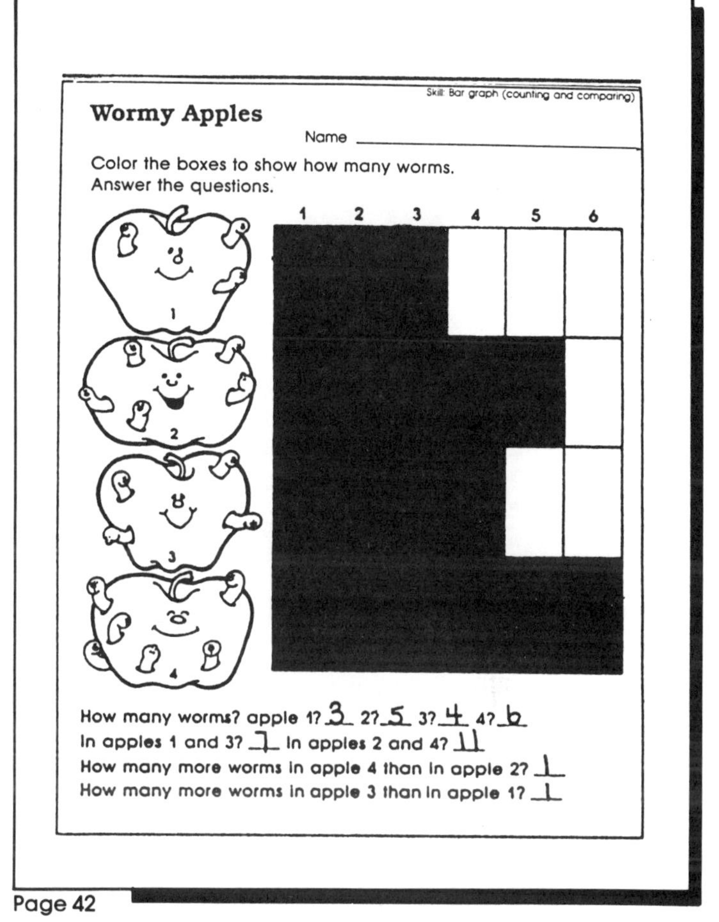

Page 42

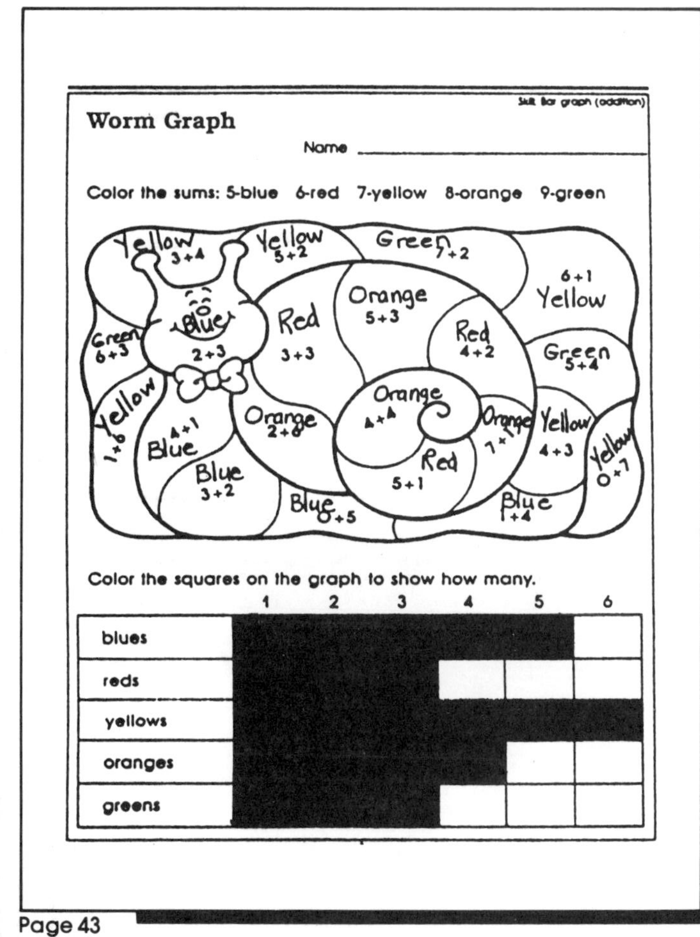

Page 43

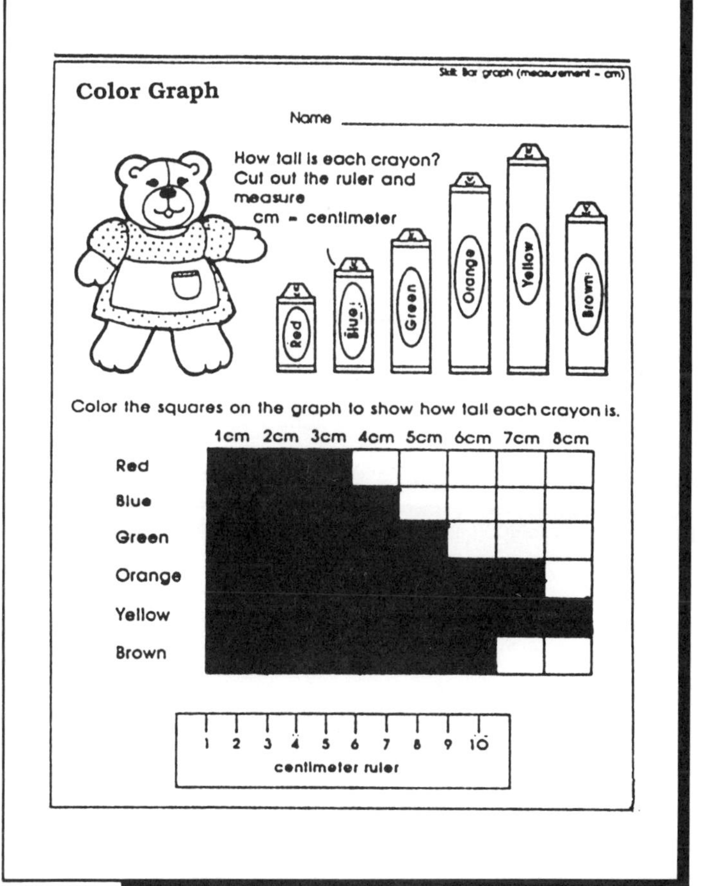

Page 44

Answer Key

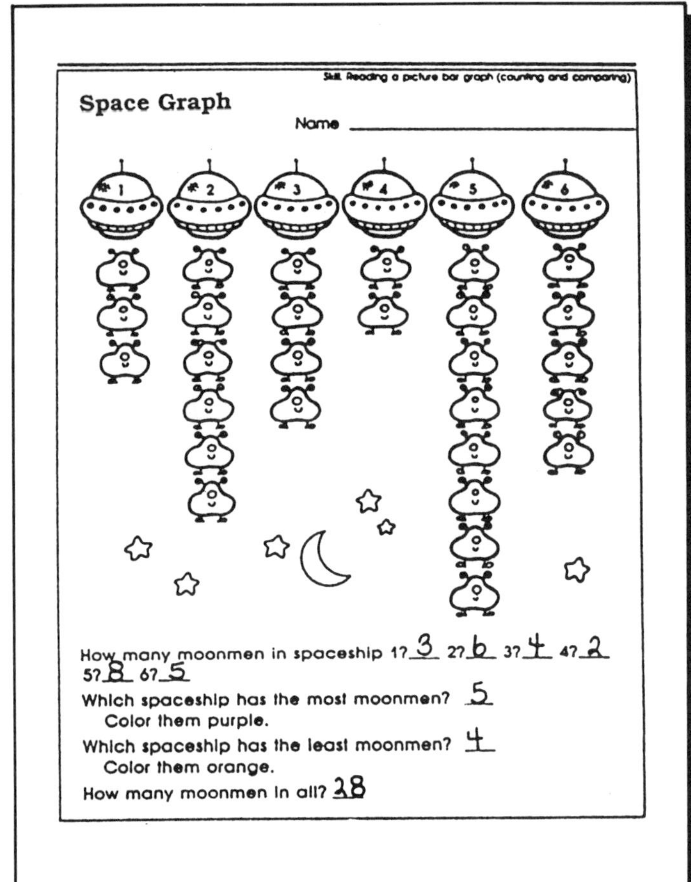

Skill: Reading a picture bar graph (counting and comparing)

Space Graph

Name ____________

How many moonmen in spaceship 1? 3 2? 6 3? 4 4? 2
5? 8 6? 5

Which spaceship has the most moonmen? 5
Color them purple.

Which spaceship has the least moonmen? 4
Color them orange.

How many moonmen in all? 28

Page 45

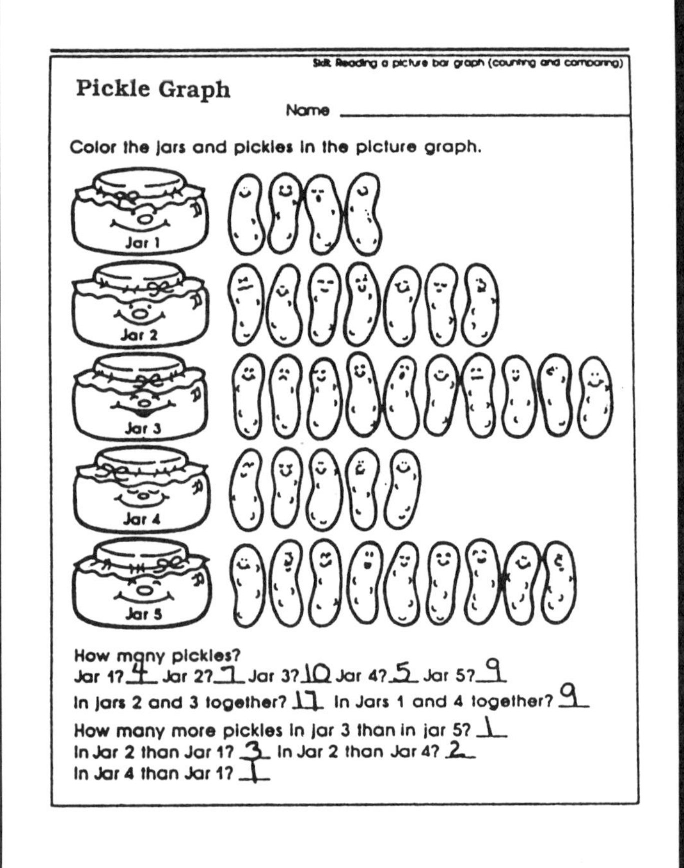

Skill: Reading a picture bar graph (counting and comparing)

Pickle Graph

Name ____________

Color the jars and pickles in the picture graph.

Jar 1
Jar 2
Jar 3
Jar 4
Jar 5

How many pickles?
Jar 1? 4 Jar 2? 7 Jar 3? 10 Jar 4? 5 Jar 5? 9
In jars 2 and 3 together? 17 In Jars 1 and 4 together? 9
How many more pickles in jar 3 than in jar 5? 1
In Jar 2 than Jar 1? 3 In Jar 2 than Jar 4? 2
In Jar 4 than Jar 1? 1

Page 46

Skill: Reading a picture bar graph (counting)

Pat's Fish

Name ____________

This picture graph shows how many fish Pat caught.

First Saturday — blue, yellow
Second Saturday
Third Saturday — red
Fourth Saturday — green

Color the fish Pat caught on the third Saturday red.
Color the fish he caught on the first Saturday blue,
the second Saturday yellow, and the fourth Saturday green.
How many fish did he catch on the first Saturday? 3
second Saturday? 6 third Saturday? 4 fourth Saturday? 2

Page 47

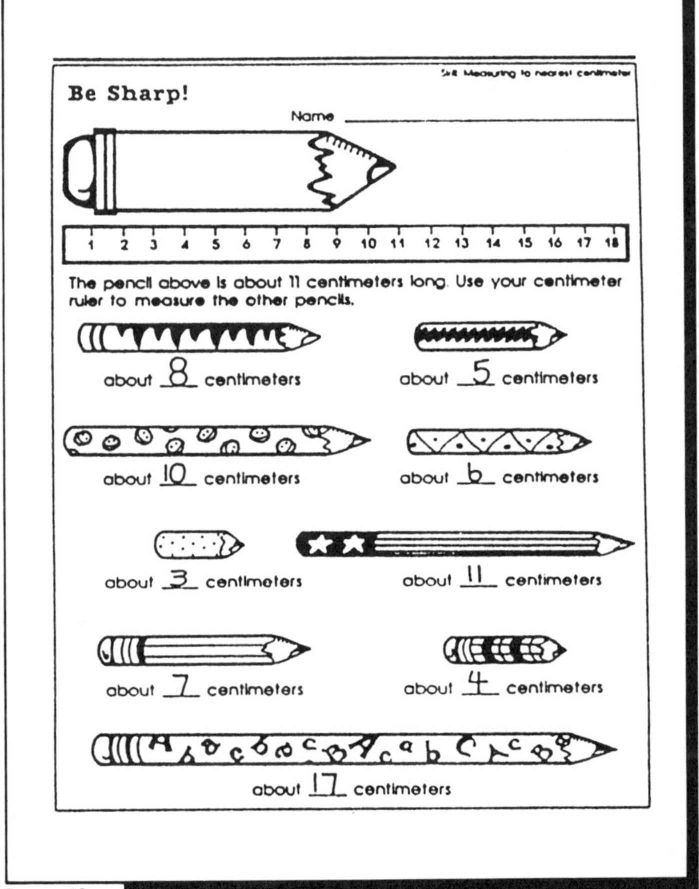

Skill: Measuring to nearest centimeter

Be Sharp!

Name ____________

1 2 3 4 5 6 7 8 9 10 11 12 13 14 15 16 17 18

The pencil above is about 11 centimeters long. Use your centimeter ruler to measure the other pencils.

about 8 centimeters about 5 centimeters

about 10 centimeters about 6 centimeters

about 3 centimeters about 11 centimeters

about 7 centimeters about 4 centimeters

about 17 centimeters

Page 48

Answer Key

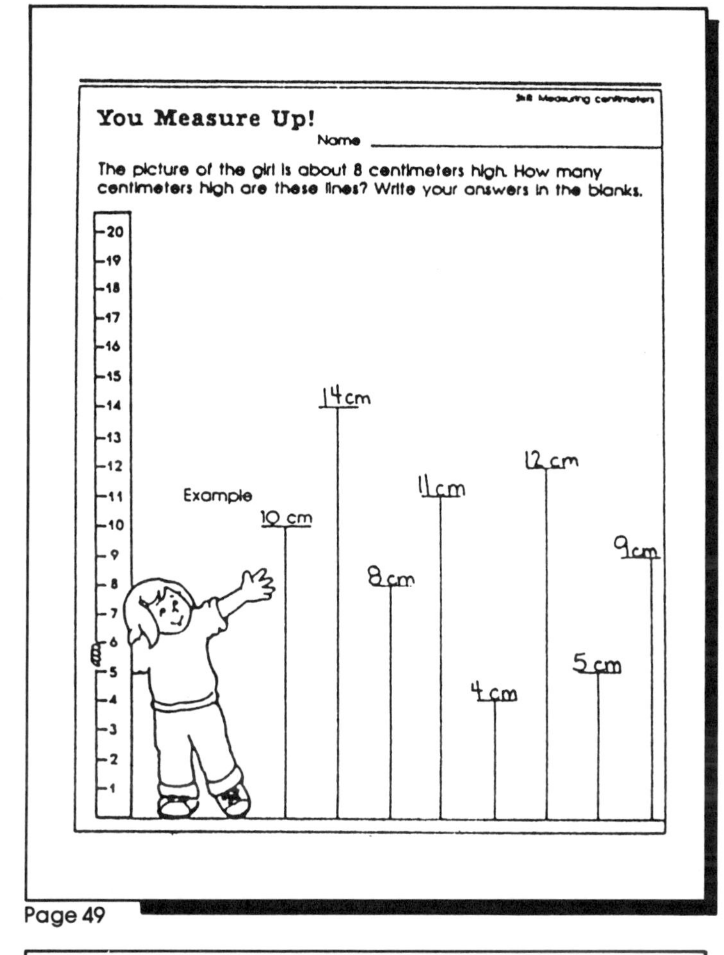

Page 49

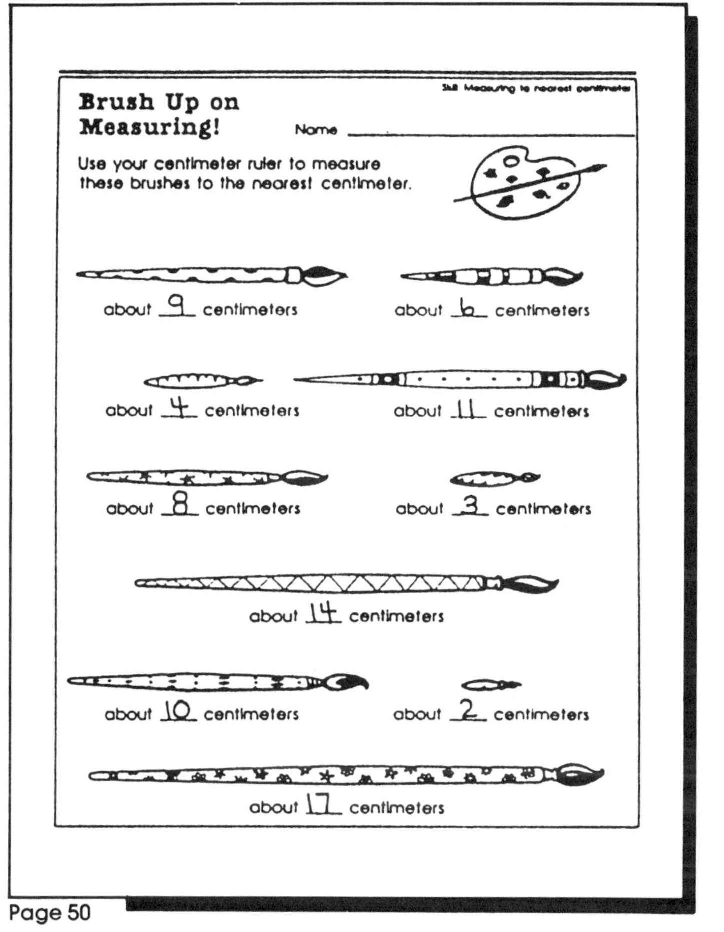

Page 50

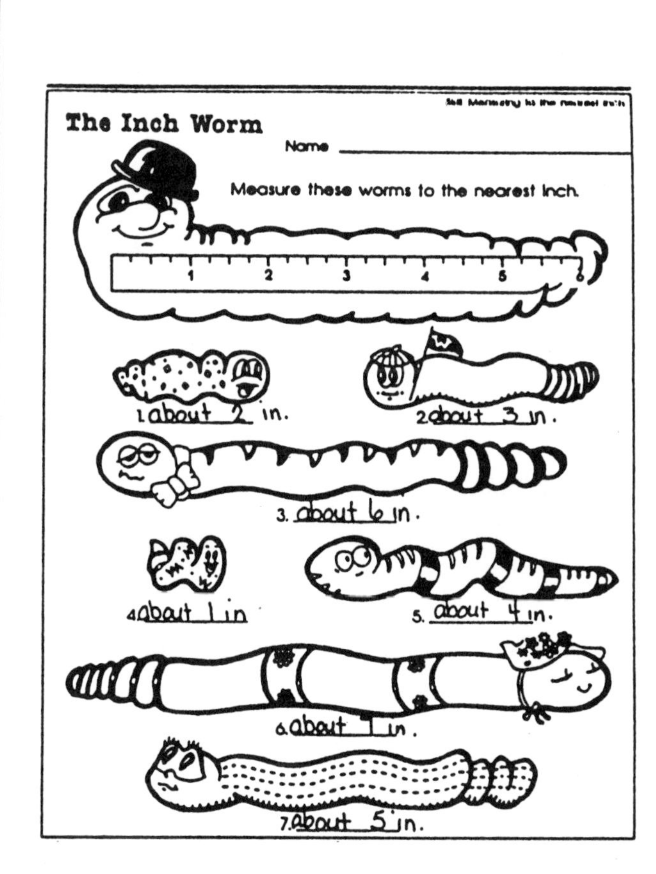

The Inch Worm
Name
Measure these worms to the nearest inch.
1 2 3 4 5 6
1. about 2 in.
2. about 3 in.
3. about 6 in.
4. about 1 in
5. about 4 in.
6. about 7 in.
7. about 5 in.

Page 51

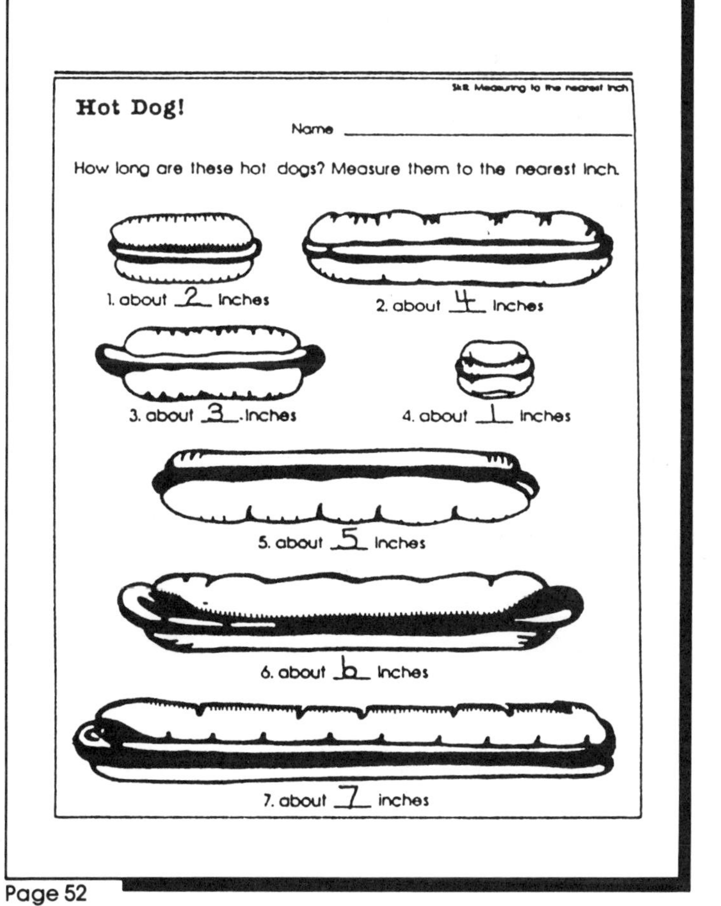

Skill: Measuring to the nearest inch
Hot Dog!
Name
How long are these hot dogs? Measure them to the nearest inch.
1. about 2 inches
2. about 4 inches
3. about 3 inches
4. about 1 inches
5. about 5 inches
6. about 6 inches
7. about 7 inches

Page 52

Answer Key

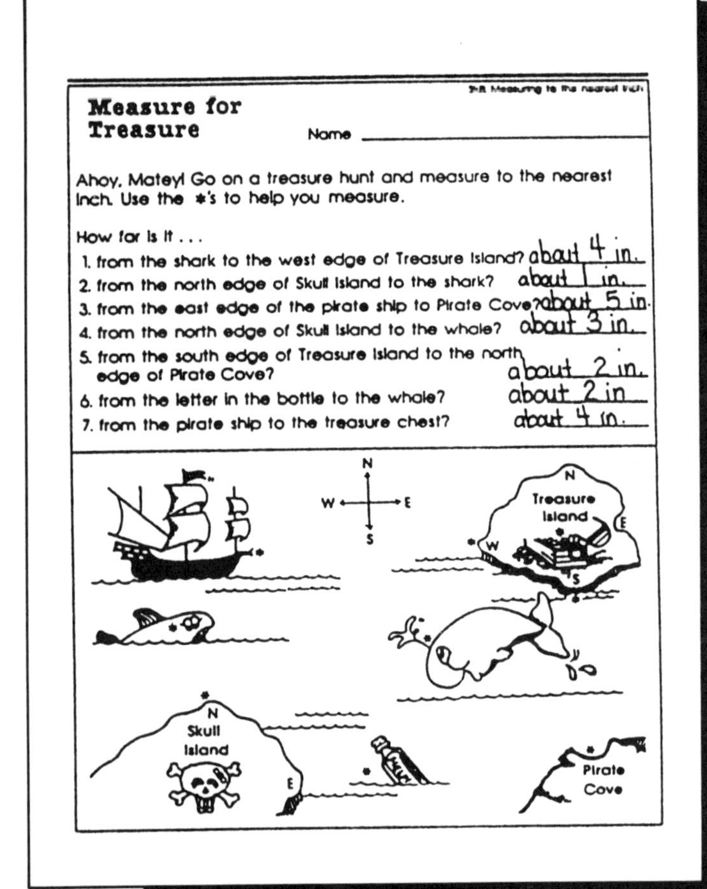

Skill: Measuring to the nearest inch

Measure for Treasure

Name ______________

Ahoy, Matey! Go on a treasure hunt and measure to the nearest inch. Use the *'s to help you measure.

How far is it . . .

1. from the shark to the west edge of Treasure Island? about 4 in.
2. from the north edge of Skull Island to the shark? about 1 in.
3. from the east edge of the pirate ship to Pirate Cove? about 5 in.
4. from the north edge of Skull Island to the whale? about 3 in.
5. from the south edge of Treasure Island to the north edge of Pirate Cove? about 2 in.
6. from the letter in the bottle to the whale? about 2 in
7. from the pirate ship to the treasure chest? about 4 in.

Page 53

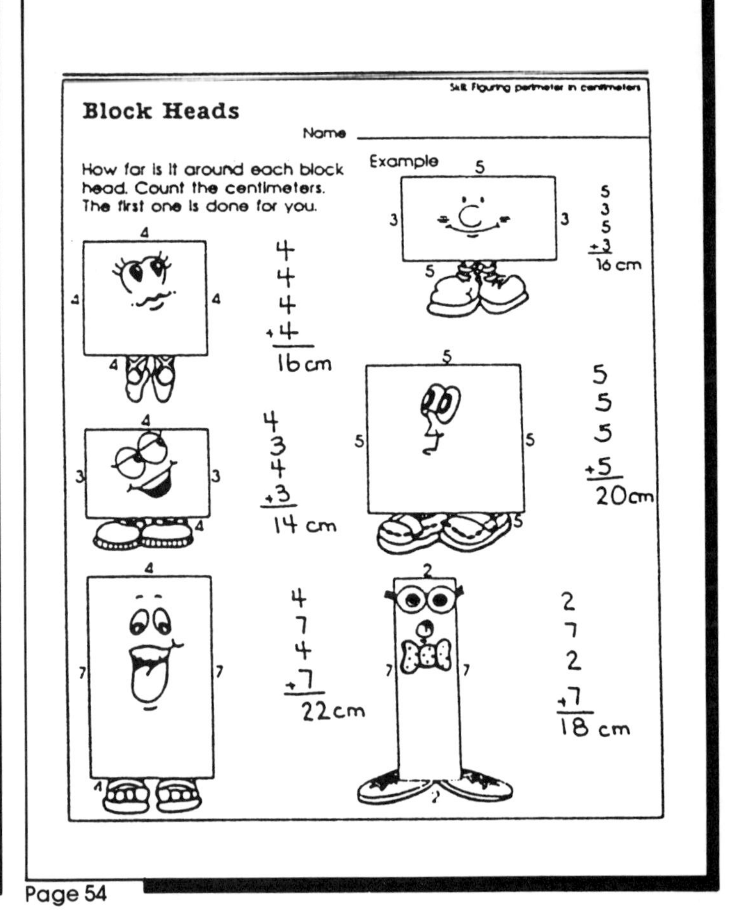

Skill: Figuring perimeter in centimeters

Block Heads

Name ______________

How far is it around each block head. Count the centimeters. The first one is done for you.

Page 54

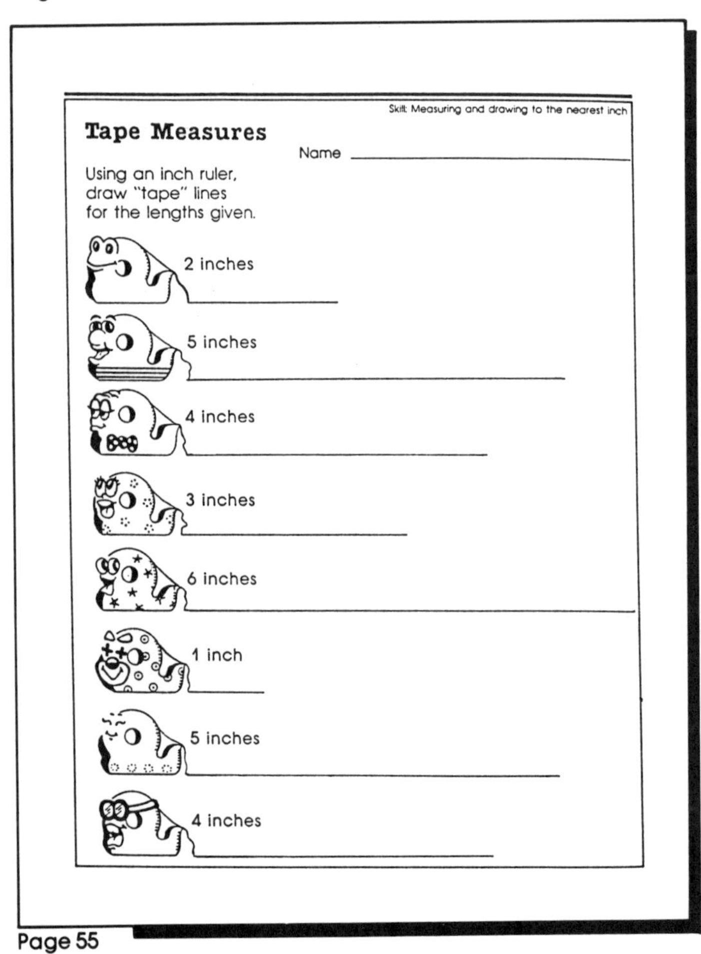

Skill: Measuring and drawing to the nearest inch

Tape Measures

Name ______________

Using an inch ruler, draw "tape" lines for the lengths given.

- 2 inches
- 5 inches
- 4 inches
- 3 inches
- 6 inches
- 1 inch
- 5 inches
- 4 inches

Page 55

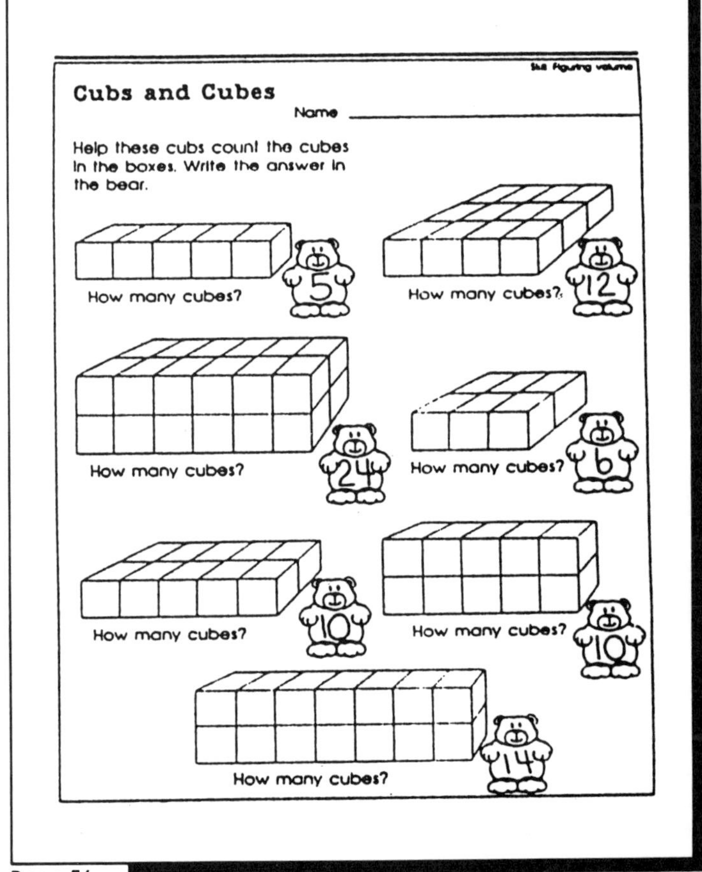

Skill: Figuring volume

Cubs and Cubes

Name ______________

Help these cubs count the cubes in the boxes. Write the answer in the bear.

How many cubes? 5

How many cubes? 12

How many cubes? 24

How many cubes? 6

How many cubes? 10

How many cubes? 10

How many cubes? 14

Page 56

Answer Key

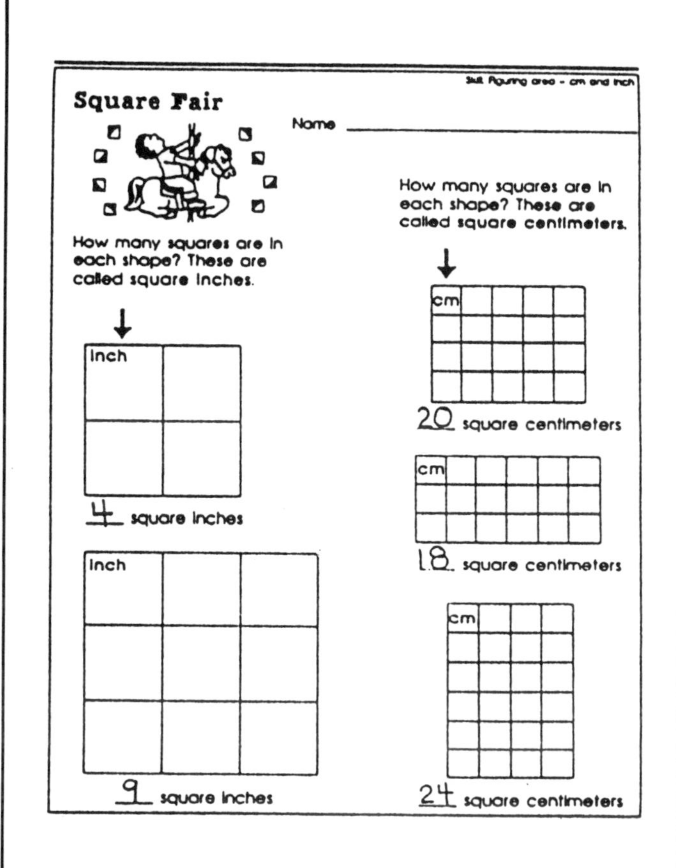
Skill: Figuring area - cm and inch

Square Fair

Name

How many squares are in each shape? These are called square inches.

inch

4 square inches

inch

9 square inches

How many squares are in each shape? These are called square centimeters.

cm

20 square centimeters

cm

18 square centimeters

cm

24 square centimeters

Page 57

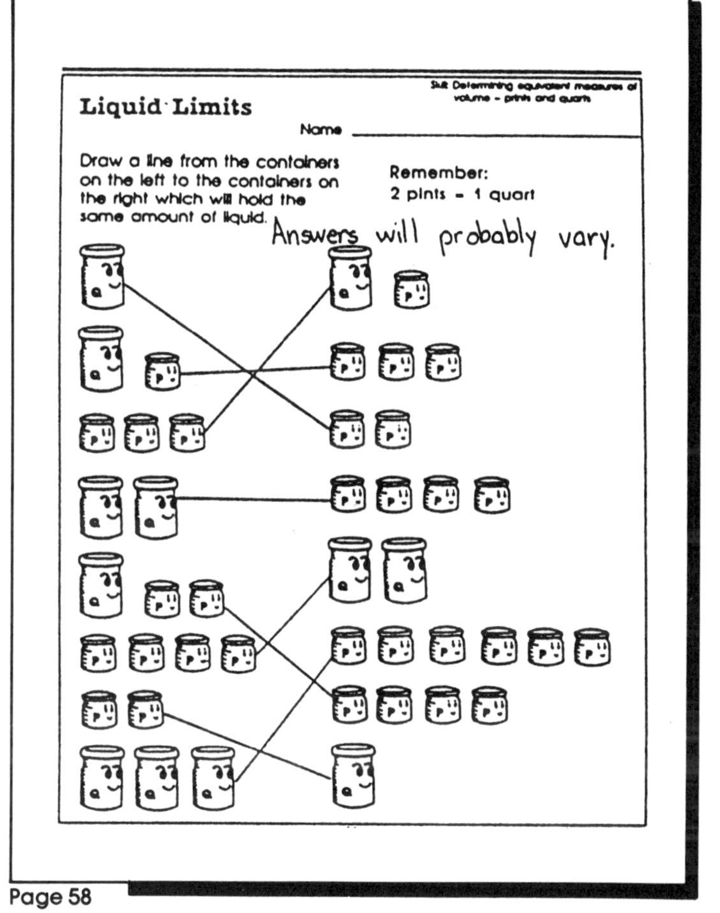
Skill: Determining equivalent measures of volume - pints and quarts

Liquid Limits

Name

Draw a line from the containers on the left to the containers on the right which will hold the same amount of liquid.

Remember:
2 pints = 1 quart

Answers will probably vary.

Page 58

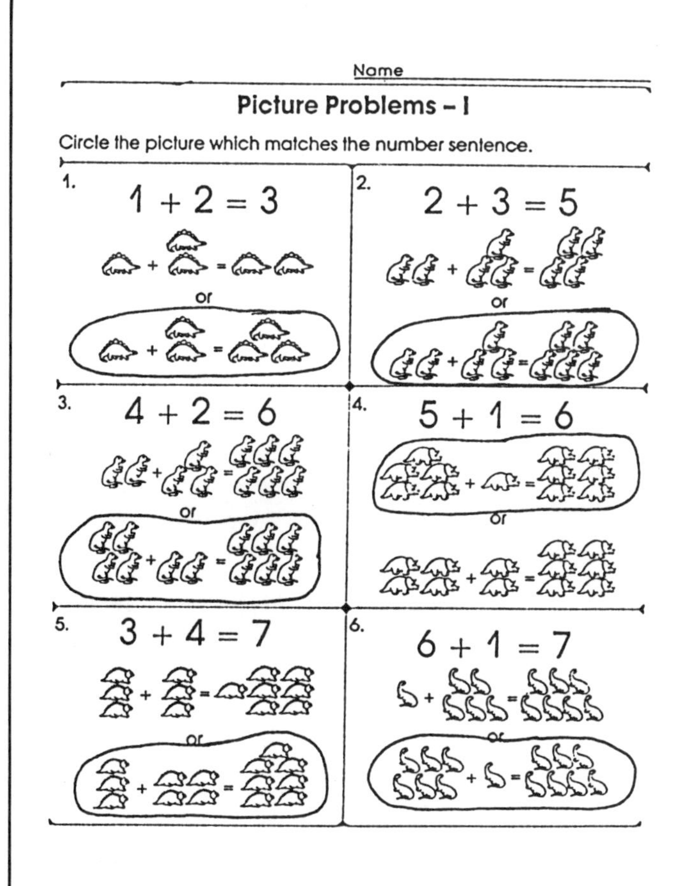
Name

Picture Problems – I

Circle the picture which matches the number sentence.

1. $1 + 2 = 3$

2. $2 + 3 = 5$

3. $4 + 2 = 6$

4. $5 + 1 = 6$

5. $3 + 4 = 7$

6. $6 + 1 = 7$

Page 59

Name

Picture Problems – II

Look at the pictures and finish the number sentences.

1. $2 + 3 = 5$

2. $1 + 7 = 8$

3. $4 + 3 = 7$

4. $5 + 0 = 5$

5. $3 + 3 = 6$

6. $4 + 5 = 9$

Page 60

Answer Key

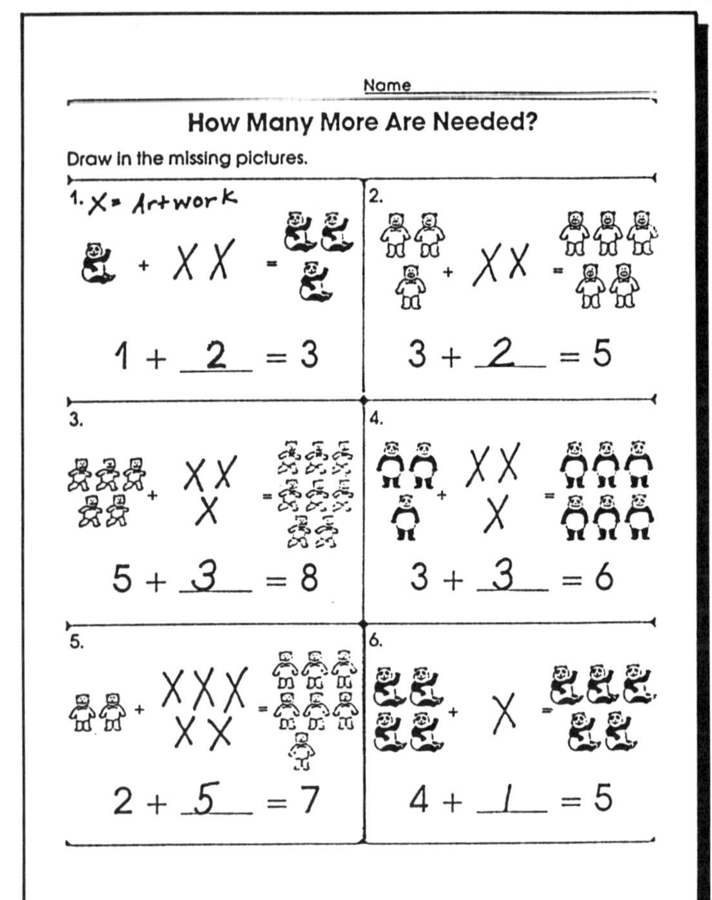

Page 61

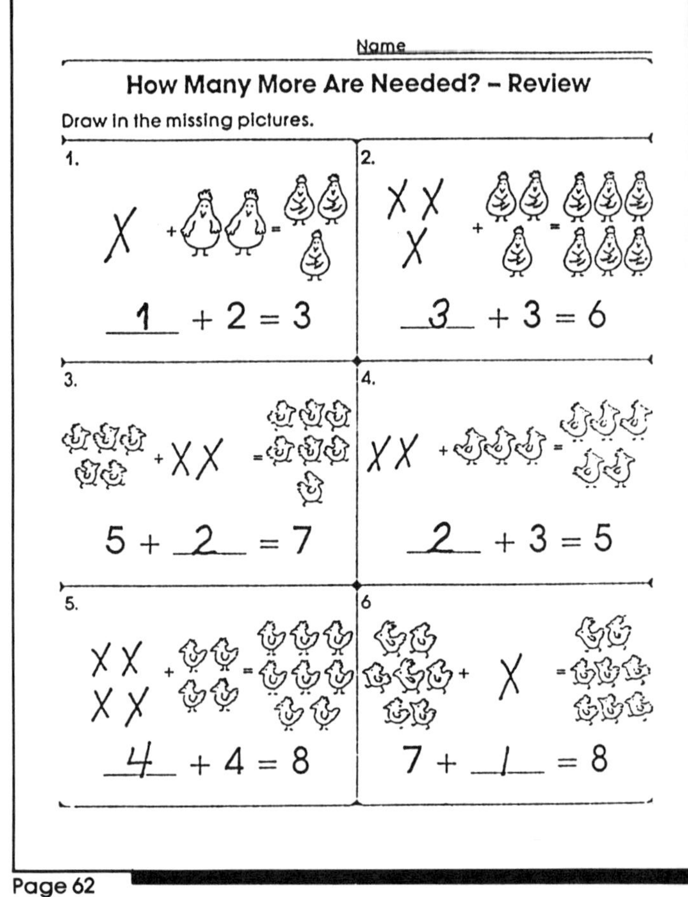

Page 62

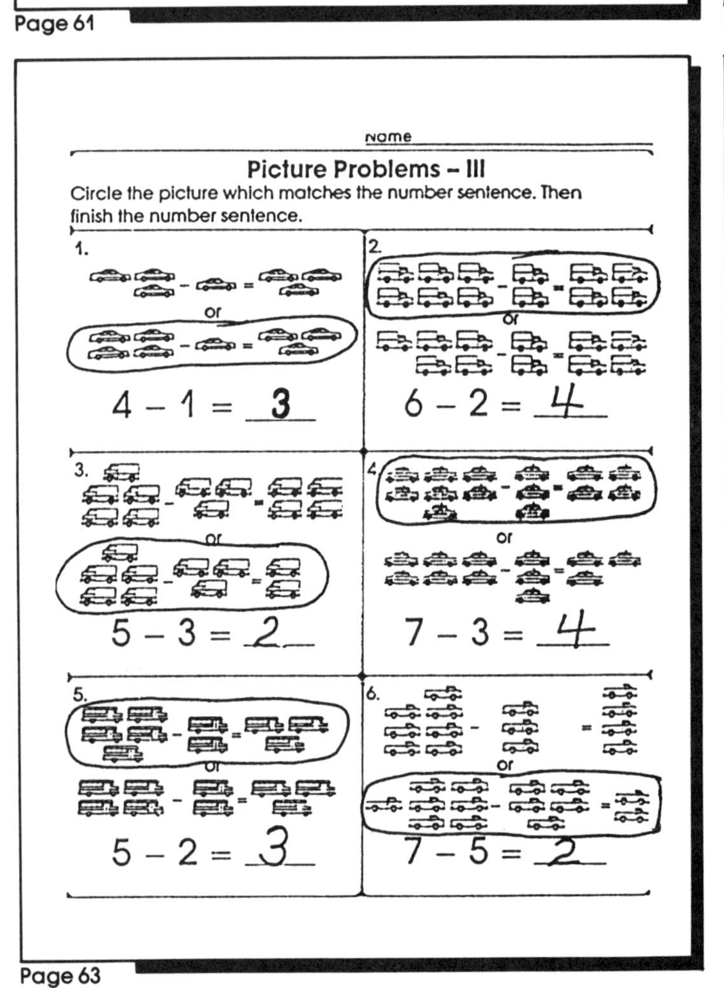

Page 63

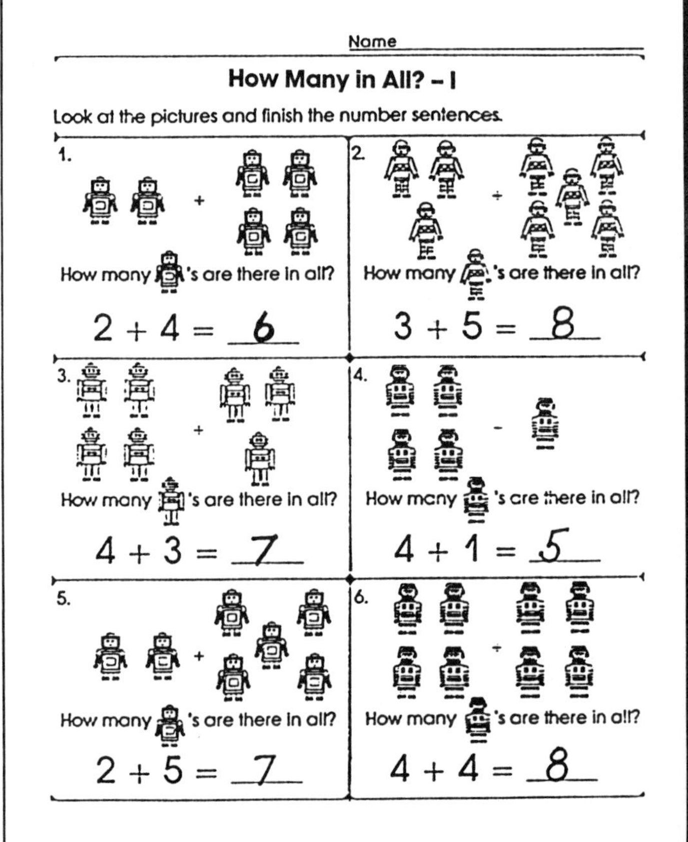

Page 64

Answer Key

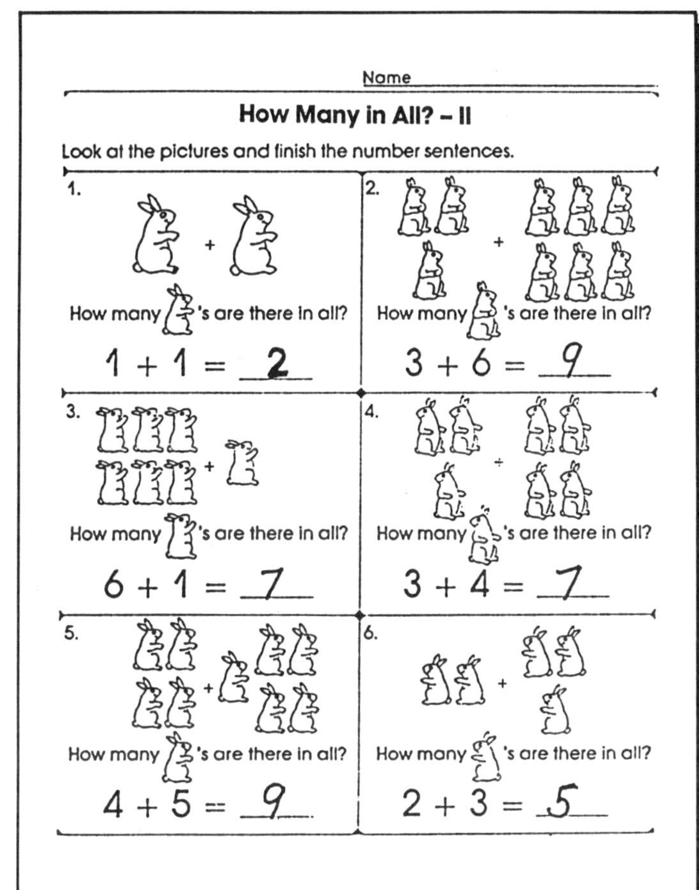
Name

How Many in All? – II

Look at the pictures and finish the number sentences.

1. How many 's are there in all? 1 + 1 = 2
2. How many 's are there in all? 3 + 6 = 9
3. How many 's are there in all? 6 + 1 = 7
4. How many 's are there in all? 3 + 4 = 7
5. How many 's are there in all? 4 + 5 = 9
6. How many 's are there in all? 2 + 3 = 5

Page 65

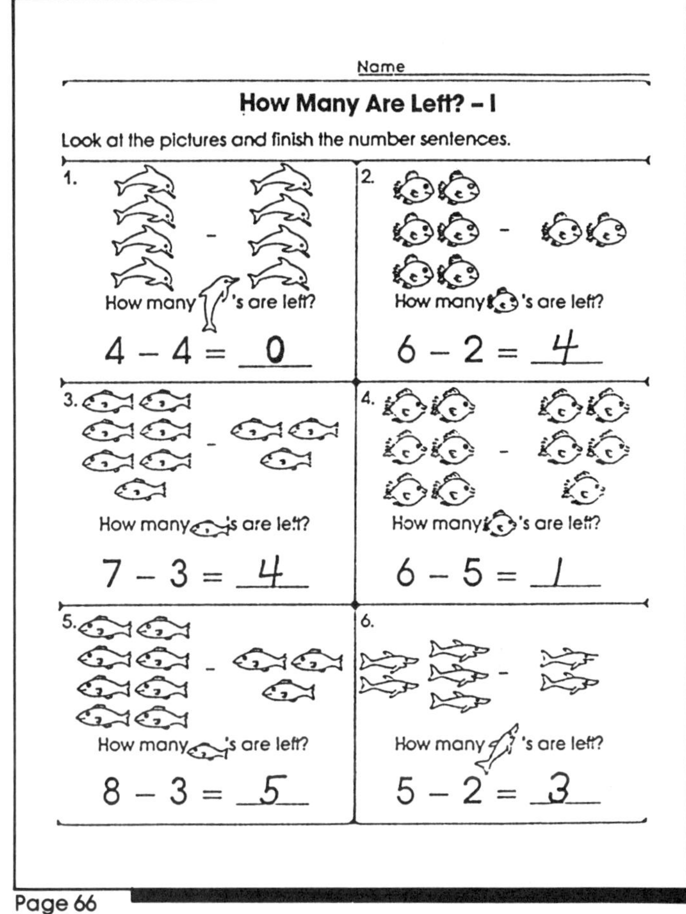
Name

How Many Are Left? – I

Look at the pictures and finish the number sentences.

1. How many 's are left? 4 – 4 = 0
2. How many 's are left? 6 – 2 = 4
3. How many 's are left? 7 – 3 = 4
4. How many 's are left? 6 – 5 = 1
5. How many 's are left? 8 – 3 = 5
6. How many 's are left? 5 – 2 = 3

Page 66

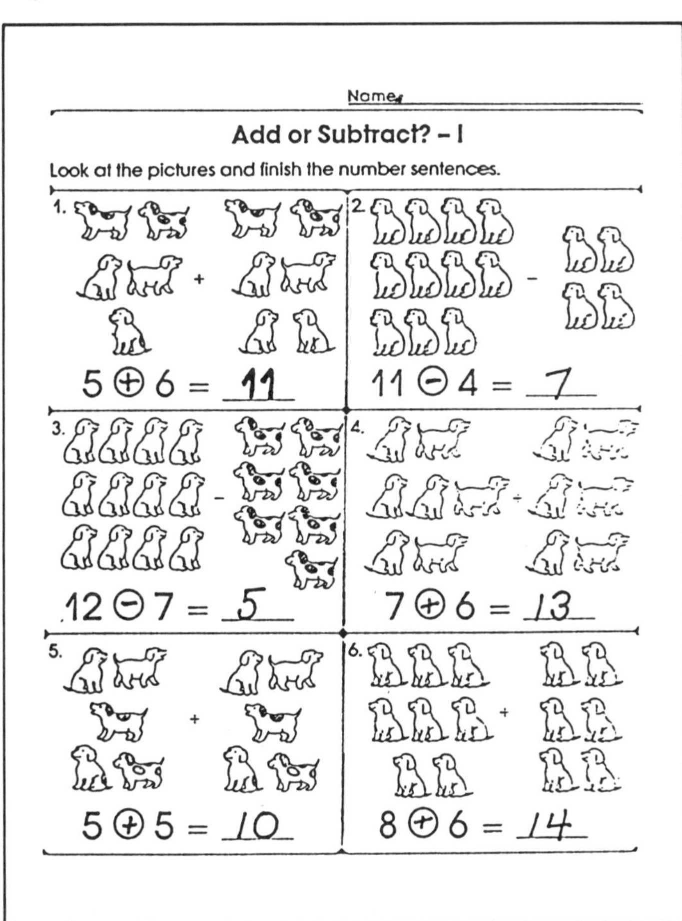
Name

Add or Subtract? – I

Look at the pictures and finish the number sentences.

1. 5 ⊕ 6 = 11
2. 11 ⊖ 4 = 7
3. 12 ⊖ 7 = 5
4. 7 ⊕ 6 = 13
5. 5 ⊕ 5 = 10
6. 8 ⊕ 6 = 14

Page 67

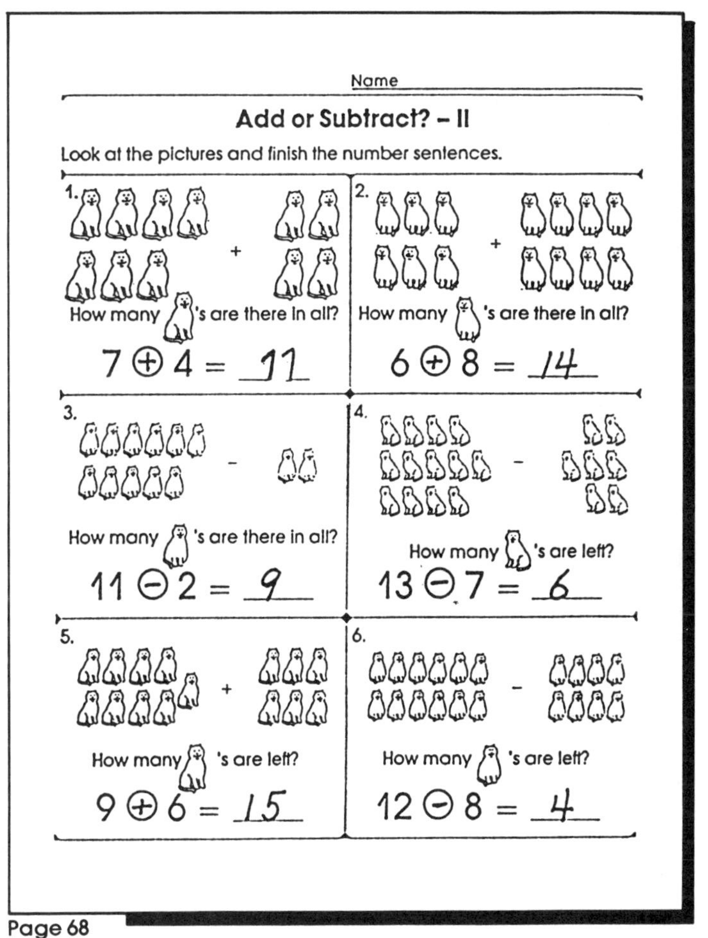
Name

Add or Subtract? – II

Look at the pictures and finish the number sentences.

1. How many 's are there in all? 7 ⊕ 4 = 11
2. How many 's are there in all? 6 ⊕ 8 = 14
3. How many 's are there in all? 11 ⊖ 2 = 9
4. How many 's are left? 13 ⊖ 7 = 6
5. How many 's are left? 9 ⊕ 6 = 15
6. How many 's are left? 12 ⊖ 8 = 4

Page 68

Answer Key

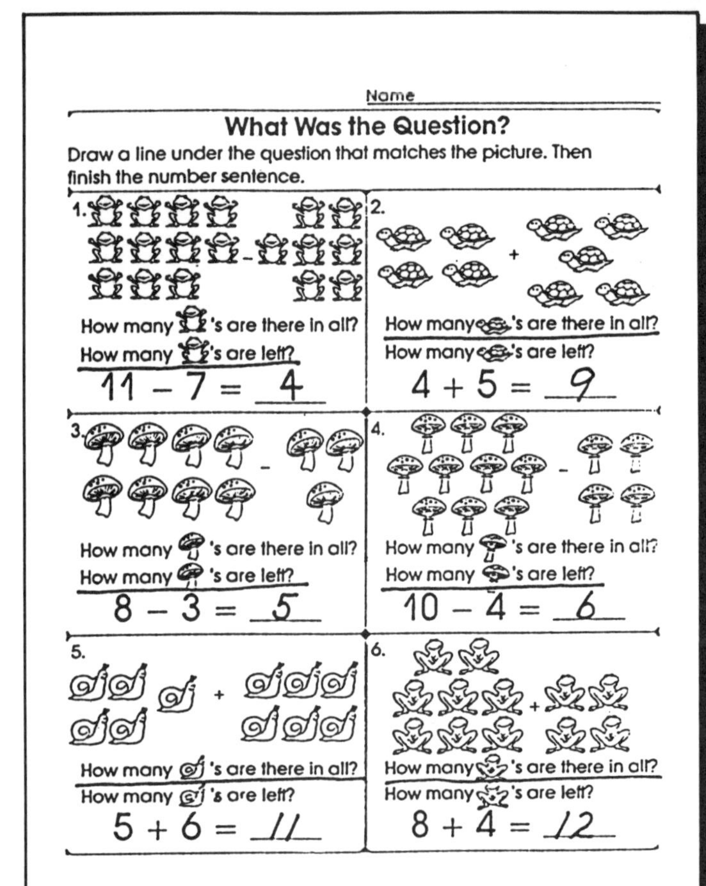

Name

What Was the Question?

Draw a line under the question that matches the picture. Then finish the number sentence.

1. How many 's are there in all?
How many 's are left?
11 – 7 = 4

2. How many 's are there in all?
How many 's are left?
4 + 5 = 9

3. How many 's are there in all?
How many 's are left?
8 – 3 = 5

4. How many 's are there in all?
How many 's are left?
10 – 4 = 6

5. How many 's are there in all?
How many 's are left?
5 + 6 = 11

6. How many 's are there in all?
How many 's are left?
8 + 4 = 12

Page 69

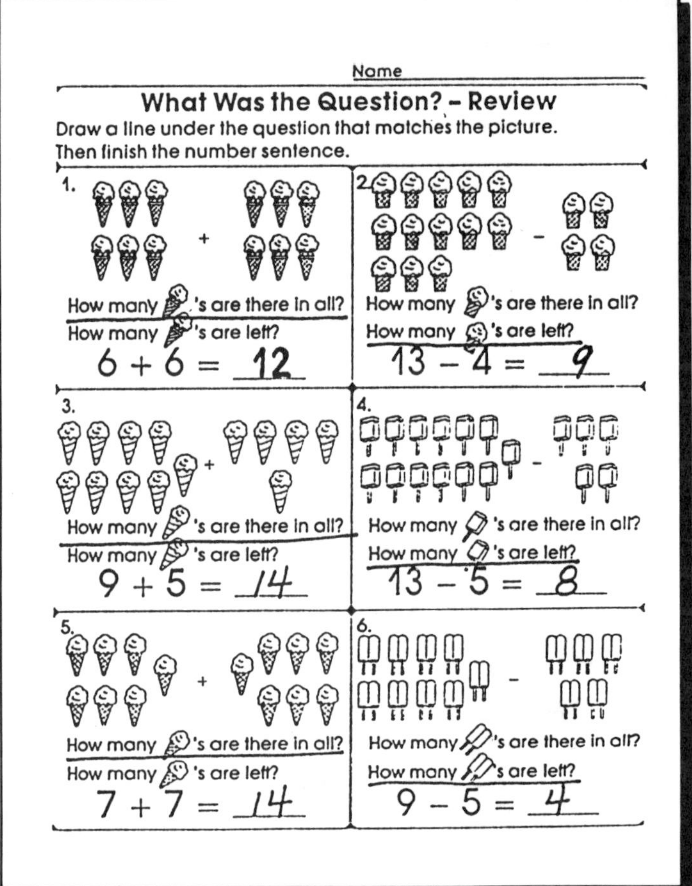

Name

What Was the Question? – Review

Draw a line under the question that matches the picture. Then finish the number sentence.

1. How many 's are there in all?
How many 's are left?
6 + 6 = 12

2. How many 's are there in all?
How many 's are left?
13 – 4 = 9

3. How many 's are there in all?
How many 's are left?
9 + 5 = 14

4. How many 's are there in all?
How many 's are left?
13 – 5 = 8

5. How many 's are there in all?
How many 's are left?
7 + 7 = 14

6. How many 's are there in all?
How many 's are left?
9 – 5 = 4

Page 70

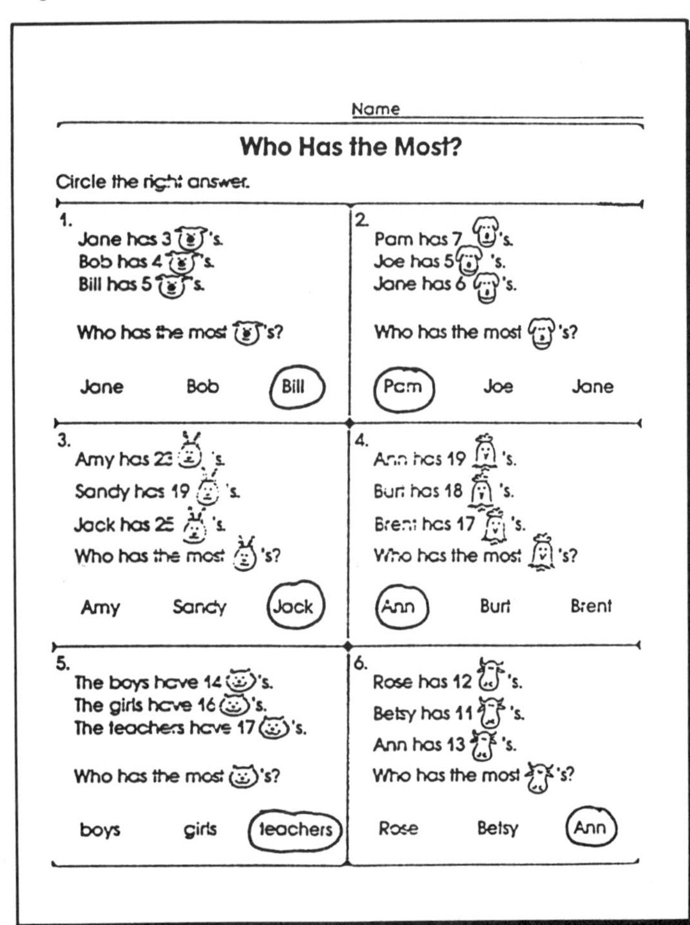

Name

Who Has the Most?

Circle the right answer.

1. Jane has 3 's.
Bob has 4 's.
Bill has 5 's.
Who has the most 's?
Jane Bob (Bill)

2. Pam has 7 's.
Joe has 5 's.
Jane has 6 's.
Who has the most 's?
(Pam) Joe Jane

3. Amy has 23 's.
Sandy has 19 's.
Jack has 25 's.
Who has the most 's?
Amy Sandy (Jack)

4. Ann has 19 's.
Burt has 18 's.
Brent has 17 's.
Who has the most 's?
(Ann) Burt Brent

5. The boys have 14 's.
The girls have 16 's.
The teachers have 17 's.
Who has the most 's?
boys girls (teachers)

6. Rose has 12 's.
Betsy has 11 's.
Ann has 13 's.
Who has the most 's?
Rose Betsy (Ann)

Page 71

Name

Who Has the Least?

Circle the right answer.

1. Pat had 4 's.
Charles had 3 's.
Jane had 5 's.
Who had the least 's?
Pat (Charles) Jane

2. Jeff has 5 's.
John has 4 's.
Bill has 6 's.
Who has the least number of 's?
Jeff (John) Bill

3. Jane has 7 's.
Peg has 9 's.
Fred has 8 's.
Who has the least number of 's?
(Jane) Peg Fred

4. Charles bought 12 's.
Rose bought 6 's.
Mother bought 24 's.
Who bought the least number of 's?
Charles (Rose) Mother

5. John had 9 's.
Jack had 8 's.
Jeff had 7 's.
Who had the least number of 's?
John Jack (Jeff)

6. Alma bought 12 's.
Nina bought 16 's.
Marty bought 13 's.
Who bought the least number of 's?
(Alma) Nina Marty

Page 72

Answer Key

Page 73

Name ____________

How Many in All? III

The key words in all tell you to add. Circle the key words in all and solve the problems.

1. Jack has 4 white shirts and 2 yellow shirts. How many shirts does Jack have in all?

   $4 + 2 = 6$

2. Joan has 4 pink blouses and 6 red ones. How many blouses does Joan have in all?

   $4 + 6 = 10$

3. Mack has 3 pairs of summer pants and 8 pairs of winter pants. How many pairs of pants does Mack have in all?

   $3 + 8 = 11$

4. Betsy has 2 black skirts and 7 blue skirts. In all how many skirts does Betsy have?

   $2 + 7 = 9$

5. Willis has 5 knit hats and 5 cloth hats. How many hats does Willis have in all?

   $5 + 5 = 10$

Page 74

Name ____________

How Many in All? IV

Circle the addition key words in all and solve the problems.

1. On the block where Cindy lives there are 7 brick houses and 5 stone houses. How many houses are there in all?

   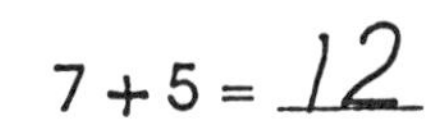

   $7 + 5 = 12$

2. One block from Cindy's house there are 7 white houses and 4 grey houses. How many houses are there in all?

   $7 + 4 = 11$

3. Near Cindy's house there are 3 grocery stores and 5 discount stores. How many stores are there in all?

   $3 + 5 = 8$

4. Children live in 8 of the two-story houses, and children live in 2 of the one-story houses. How many houses in all have children living in them?

   $8 + 2 = 10$

5. In Cindy's neighborhood 4 students are in high school and 9 are in elementary school. In all, how many children are in school?

   $4 + 9 = 13$

Page 75

Name ____________

How Many in All? V

Circle the addition key words in all and solve the problems.

1. At the park there are 3 baseball games and 6 basketball games being played. How many games are being played in all?

   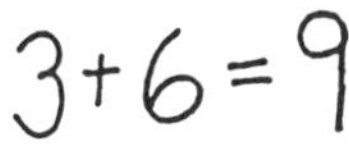

   $3 + 6 = 9$

2. In the park 9 mothers are pushing their babies in strollers, and 8 are carrying their babies in baskets. How many mothers in all have their babies with them in the park?

   $9 + 8 = 17$

3. On one team there are 6 boys and 3 girls. How many team members are there in all?

   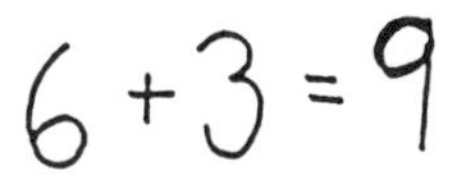

   $6 + 3 = 9$

4. At one time there were 8 men and 4 boys pitching horseshoes. In all how many people were pitching horseshoes?

   $8 + 4 = 12$

5. While playing basketball, 4 of the players were wearing gym shoes and 6 were not. How many basketball players were there in all?

   $4 + 6 = 10$

Page 76

Name ____________

How Many Are Left? – II

The key word left tells you to subtract. Circle the key word left and solve the problems.

1. Bill had 10 kittens, but 4 of them ran away. How many kittens does he have left?

   $10 - 4 = 6$

2. There were 12 rabbits eating clover. Dogs chased 3 of them away. How many rabbits were left?

   $12 - 3 = 9$

3. Bill saw 11 birds eating from the bird feeders in his back yard. A cat scared 7 of them away. How many birds were left at the feeders?

   $11 - 7 = 4$

4. There were 14 frogs on the bank of the pond. Then 9 of them hopped into the water. How many frogs were left on the bank?

   $14 - 9 = 5$

5. Bill counted 15 robins in his yard. Then 8 of the robins flew away. How many robins were left in the yard?

   $15 - 8 = 7$

Answer Key

Name

How Many Are Left? III

Circle the subtraction key word left and solve the problems.

1. In Maggy's classroom there are 12 girls. One day 4 of the girls went home with the flu. How many girls were (left) in school that day?

$12 - 4 = 8$

2. Maggy is in 10 different clubs. This week 5 of them will not meet. How many of Maggy's clubs are (left) to meet this week?

$10 - 5 = 5$

3. Maggy had 16 crayons. She broke 9 of them. How many crayons does Maggy have (left)?

$16 - 9 = 7$

4. There are 13 boys in Maggy's classroom. One morning 8 of the boys went to the gym. How many were (left) in the classroom?

$13 - 8 = 5$

5. One day 4 of the 13 boys were called in from the playground. How many of the boys were (left) on the playground?

$13 - 4 = 9$

Page 77

Name

Add or Subtract? – III

The key words in all tell you to add. The key word left tells you to subtract. Circle the key words and solve the problems.

1. The pet store has 3 large dogs and 5 small dogs. How many dogs are there (in all)?

$3 \oplus 5 = 8$

2. The pet store had 9 parrots and then sold 4 of them. How many parrots does the pet store have (left)?

$9 \ominus 4 = 5$

3. The pet store gave Linda's class 2 adult gerbils and 9 young ones. How many gerbils did Linda's class get (in all)?

$2 \oplus 9 = 11$

4. At the pet store 3 of the 8 myna birds were sold. How many myna birds are (left) in the pet store?

$8 \ominus 3 = 5$

5. The monkey at the pet store has 5 rubber toys and 4 wooden toys. How many toys does it have (in all)?

$5 \oplus 4 = 9$

Page 78

Name

Add or Subtract? – IV

Circle the key words in all and left and solve the problems.

1. Alex picked 5 tomatoes, and Don picked 4. How many tomatoes did they pick (in all)?

$5 + 4 = 9$

2. George picked 17 apples and ate 6 of them. How many apples does he have (left)?

$17 - 6 = 11$

3. Mary picked 4 peppers, and Sue picked 9 peppers. How many peppers (in all) did they pick?

$4 + 9 = 13$

4. There were 11 roses on the bush. Cindy picked 5 of them. How many roses were (left) on the bush?

$11 - 5 = 6$

5. Max picked 8 bunches of grapes. Joe picked 6 bunches. How many bunches of grapes were picked (in all)?

$8 + 6 = 14$

Page 79

Name

Key Word Practice – I

The key word altogether tells you to add. They key words many more tell you to subtract. Circle the key words and solve the problems.

1. Judy has 9 pennies. Mack has 3 pennies. How (many more) pennies does Judy have than Mack?

$9 - 3 = 6$

2. Sarah has 12 dimes. Joe has 7 dimes. How (many more) dimes does Sarah have than Joe?

$12 - 7 = 5$

3. Sally was paid 8 quarters for cutting the lawn. She already had 4 quarters. How many quarters (altogether) does Sally have?

$8 + 4 = 12$

4. Jim had 5 coins. His dad gave him 9 more. How many coins does Jim have (altogether)?

$5 + 9 = 14$

5. Beth earned 2 dollars. Martha earned 11 dollars. How (many more) dollars did Martha earn than Beth?

$11 - 2 = 9$

Page 80

Answer Key

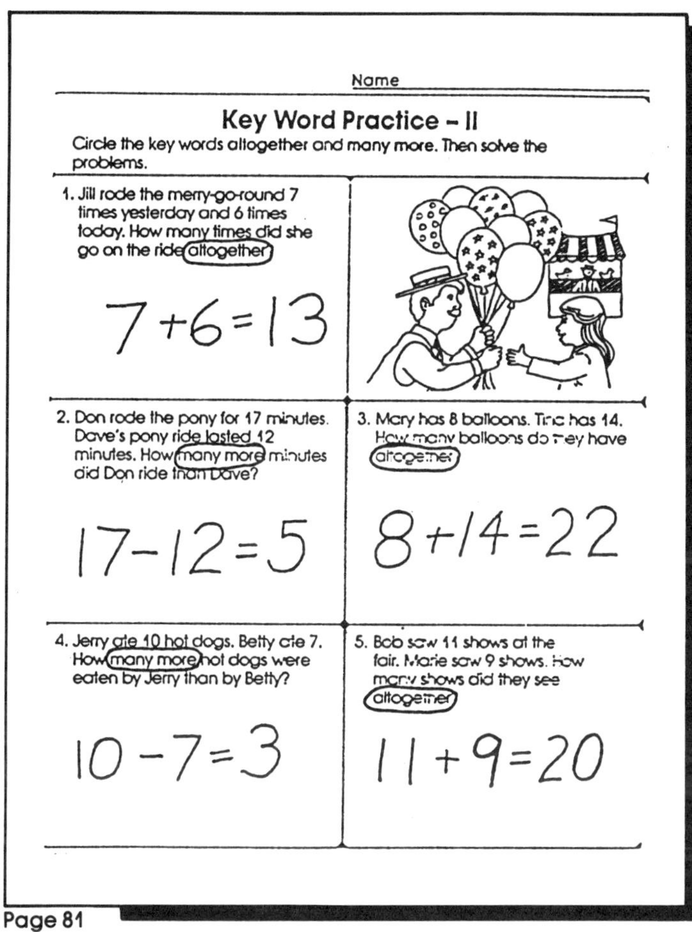

Name

Key Word Practice – II

Circle the key words altogether and many more. Then solve the problems.

1. Jill rode the merry-go-round 7 times yesterday and 6 times today. How many times did she go on the ride altogether?

   7+6=13

2. Don rode the pony for 17 minutes. Dave's pony ride lasted 12 minutes. How many more minutes did Don ride than Dave?

   17-12=5

3. Mary has 8 balloons. Tina has 14. How many balloons do they have altogether

   8+14=22

4. Jerry ate 10 hot dogs. Betty ate 7. How many more hot dogs were eaten by Jerry than by Betty?

   10-7=3

5. Bob saw 11 shows at the fair. Marie saw 9 shows. How many shows did they see altogether

   11+9=20

Page 81

Name

Fractions

Study the pictures. Then solve the story problems.

1. Bob ate 1/2 of a candy bar. Bill ate 1/4 of a candy bar. Who ate the most candy?

   Bob

2. Sue ate 3/4 of a cupcake. Angie ate 1/2 of a cupcake. Who ate the most?

   Sue

3. Pete ate 1/3 of a pie. May ate 1/4 of a pie. Who ate the most?

   Pete

4. Emily ate 1/2 of an apple. Dean ate 3/4 of an apple. Who ate the most?

   Dean

5. Amy sliced 1/4 of the cake. Sally sliced 1/3 of the cake. Who sliced the most cake?

   Sally

6. John ate 1/2 of a melon. Jeff ate 2/3 of a melon. Who ate the most?

   Jeff

Page 82

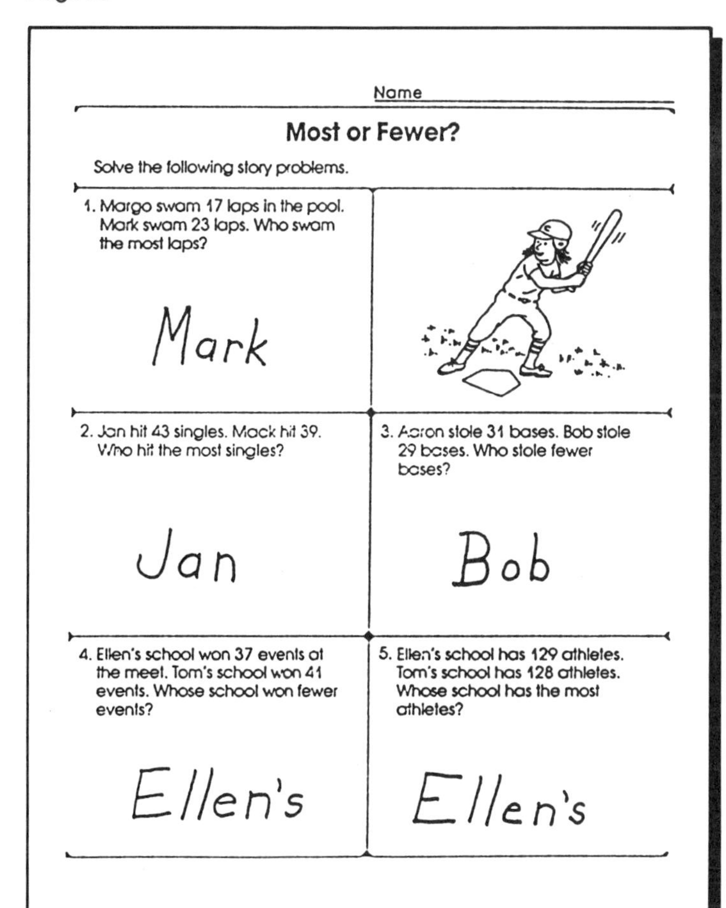

Name

Most or Fewer?

Solve the following story problems.

1. Margo swam 17 laps in the pool. Mark swam 23 laps. Who swam the most laps?

   Mark

2. Jan hit 43 singles. Mack hit 39. Who hit the most singles?

   Jan

3. Aaron stole 31 bases. Bob stole 29 bases. Who stole fewer bases?

   Bob

4. Ellen's school won 37 events at the meet. Tom's school won 41 events. Whose school won fewer events?

   Ellen's

5. Ellen's school has 129 athletes. Tom's school has 128 athletes. Whose school has the most athletes?

   Ellen's

Page 83

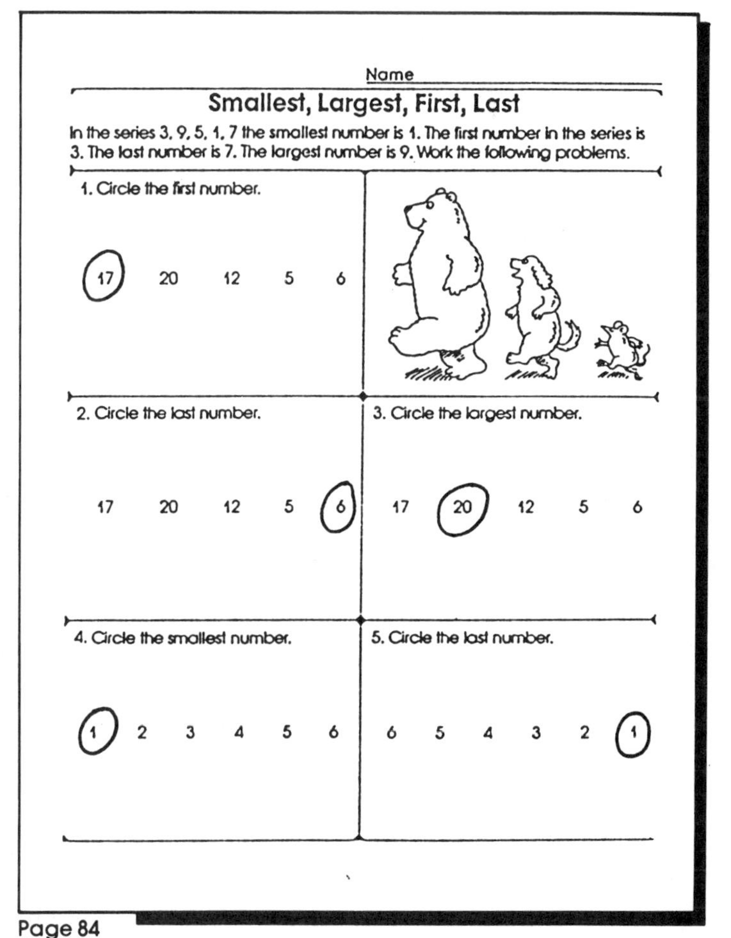

Name

Smallest, Largest, First, Last

In the series 3, 9, 5, 1, 7 the smallest number is 1. The first number in the series is 3. The last number is 7. The largest number is 9. Work the following problems.

1. Circle the first number.

   (17) 20 12 5 6

2. Circle the last number.

   17 20 12 5 (6)

3. Circle the largest number.

   17 (20) 12 5 6

4. Circle the smallest number.

   (1) 2 3 4 5 6

5. Circle the last number.

   6 5 4 3 2 (1)

Page 84

Answer Key

Name ______

Which Number Sentence? – I

Circle the number sentence that matches the story problem. Then solve the problem.

1. There are 15 men and 8 women on the police force. How many police are there on the force in all?

(15 + 8 = 23) circled

15 – 8 = ______

2. The city council has 12 members. The school board has 4 fewer members than the city council. How many members does the school board have?

12 + 4 = ______

(12 – 4 = 8) circled

3. The school board hired 12 bus drivers and 2 mechanics. How many drivers and mechanics did they hire altogether?

(12 + 2 = 14) circled

12 – 2 = ______

4. The school board hired 47 teachers and 5 administrators. How many more teachers were hired than administrators?

47 + 5 = ______

(47 – 5 = 42) circled

5. The council bought 5 fire trucks and 4 garbage trucks. How many trucks in all were bought?

(5 + 4 = 9) circled

5 – 4 = ______

Page 85

Name ______

Which Number Sentence? – II

Circle the number sentence that matches the story problem. Then solve the problem.

1. Last month 43 homeless adult dogs were kept at the pound. There were 11 fewer puppies than adult dogs at the pound. How many puppies were kept at the pound?

43 + 11 = ______

(43 – 11 = 32) circled

2. There were 12 puppies under three months old at the pound. There were 9 puppies over three months old. How many more puppies were there under three months old than over three months old?

12 + 9 = ______

(12 – 9 = 3) circled

3. Last week 8 black puppies and 7 brown puppies were given new homes. How many puppies in all were given new homes?

(8 + 7 = 15) circled

8 – 7 = ______

4. The puppies were fed 25 pounds of dog food on Monday and 9 pounds of dog food on Tuesday. How many more pounds of food did they eat on Monday than on Tuesday?

25 + 9 = ______

(25 – 9 = 16) circled

5. The dogs were given 18 gallons of water on Monday and 7 gallons of water on Tuesday. How many gallons did they drink on Monday and Tuesday altogether?

(18 + 7 = 25) circled

18 – 7 = ______

Page 86

Wacky Waldo

Skill: Simple Addition (number combinations to 18)

Name ______

Wacky Waldo is a different kind of animal trainer. He teaches mice to chase cats, he shows snakes how to fly, and he teaches fish to walk on land. He even teaches turtles to run faster than rabbits.

1. Wacky Waldo taught 8 mice to chase cats. Then he taught 9 mice to chase cats. How many mice did he teach in all?

8 + 9 = 17

2. Wacky Waldo saw 11 cats chased by his mice on Saturday. On Sunday, he saw 6 cats chased. What was the sum?

11 + 6 = 17

3. Wacky Waldo taught 9 mice to scare elephants and 4 mice to chase lions. How many mice did he teach?

9 + 4 = 13

4. Waldo found 11 whiskers on a mouse named Fred. He found 4 whiskers on a cat named Kitty. What was the total?

11 + 4 = 15

5. On Monday, Waldo taught 6 snakes to fly. On Tuesday, he taught 7 snakes to fly. How many snakes learned to fly in all?

6 + 7 = 13

6. Waldo used to have 4 flying snakes. Then he got 14 more flying snakes. How many flying snakes does Waldo have now?

4 + 14 = 18

Page 87

Ant Betty

Skill: Simple Subtraction (number combinations to 18)

Name ______

1. Ant Betty has 6 legs. Her friend Suzy Spider has 8 legs. How many more legs does Suzy have?

8 – 6 = 2

2. Ant Betty found 15 grass seeds. Mike Mouse took 6 of them. How many seeds are left?

15 – 6 = 9

3. Ant Betty has 18 brothers and 7 sisters. How many more brothers does she have?

18 – 7 = 11

4. Suzy Spider caught 13 flies last week. This week she caught 5 flies. How many more did she catch last week?

13 – 5 = 8

5. Ant Betty ate 8 seeds this week. Last week she ate 12 seeds. How many more did she eat last week?

12 – 8 = 4

6. Suzy Spider walked 17 feet in an hour. Ant Betty walked only 9 feet in an hour. How much farther did Suzy walk?

17 – 9 = 8

7. Ant Betty has 12 girl cousins. She has 17 boy cousins. How many more boy cousins does she have?

17 – 12 = 5

Page 88

Answer Key

Skill: Addition with 2 digits (no regrouping)

Ant Betty's Farm

Name ________________

1. Ant Betty has 12 brothers and 10 sisters. How many brothers and sisters does she have in all?

$$\begin{array}{r} 12 \\ +10 \\ \hline 22 \end{array}$$

2. Ant Betty picked up 15 seeds in the morning and 13 seeds in the afternoon. How many seeds did she pick up altogether?

$$\begin{array}{r} 15 \\ +13 \\ \hline 28 \end{array}$$

3. Ant Betty found 17 bits of sugar. Her brother found 11 bits of sugar. How much sugar did they find in all?

$$\begin{array}{r} 17 \\ +11 \\ \hline 28 \end{array}$$

4. Betty found 33 bread crumbs. Her friend Sue found 26 bread crumbs. How many did they find in all?

$$\begin{array}{r} 33 \\ +26 \\ \hline 59 \end{array}$$

5. Betty has 14 girl cousins and 15 boy cousins. What is the total?

$$\begin{array}{r} 14 \\ +15 \\ \hline 29 \end{array}$$

6. Ant Betty saw 23 flies in her garden and 13 flies in her house. What was the sum?

$$\begin{array}{r} 23 \\ +13 \\ \hline 36 \end{array}$$

7. Ant Betty worked 42 hours one week. The next week she worked 54 hours. What was the total?

$$\begin{array}{r} 42 \\ +54 \\ \hline 96 \end{array}$$

Page 89

Skill: Subtraction with 2 digits (no regrouping)

To Tell the Tooth

Name ________________

1. Children have 20 teeth. Adults have 32 teeth. How many more teeth do adults have?

$$\begin{array}{r} 32 \\ -20 \\ \hline 12 \end{array}$$

2. Possums have 50 teeth. Cats have 30 teeth. How many more teeth do possums have?

$$\begin{array}{r} 50 \\ -30 \\ \hline 20 \end{array}$$

3. A wild pig has 44 teeth. An aardvark has 20 teeth. What is the difference?

$$\begin{array}{r} 44 \\ -20 \\ \hline 24 \end{array}$$

4. A house mouse has 16 teeth. An anteater has 0 teeth. How many more teeth does the mouse have?

$$\begin{array}{r} 16 \\ -0 \\ \hline 16 \end{array}$$

5. An elephant has 32 teeth. A sea lion has 36 teeth. What is the difference?

$$\begin{array}{r} 36 \\ -32 \\ \hline 4 \end{array}$$

6. A hippopotamus has 36 teeth. A kangaroo has 34 teeth. What is the difference?

$$\begin{array}{r} 36 \\ -34 \\ \hline 2 \end{array}$$

7. A mongoose has 40 teeth. A possum has 50 teeth. How many more teeth does the possum have?

$$\begin{array}{r} 50 \\ -40 \\ \hline 10 \end{array}$$

Page 90

Skill: Addition with Regrouping

Wacky Waldo

Name ________________

Wacky Waldo is a great animal trainer. He teaches elephants to dance, rabbits to scare alligators and cows to fly like birds. He even teaches whales to water-ski and ride bicycles.

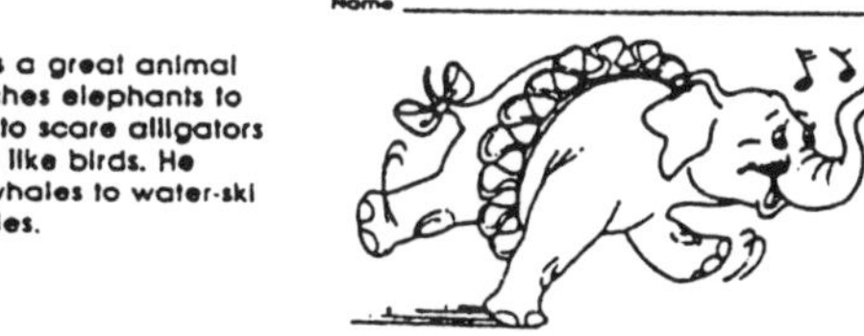

1. Waldo had 9 flying cows on Saturday and 11 flying cows on Sunday. How many flying cows did he have altogether?

$$\begin{array}{r} {}^{1}9 \\ +11 \\ \hline 20 \end{array}$$

2. Wacky Waldo taught 13 whales to water-ski and 27 whales to ride bicycles. How many whales did he teach in all?

$$\begin{array}{r} {}^{1}13 \\ +27 \\ \hline 40 \end{array}$$

3. Waldo's flying cows flew 16 miles at one time and 25 miles at another time. How many miles did they fly altogether?

$$\begin{array}{r} 16 \\ +25 \\ \hline 41 \end{array}$$

4. Waldo taught Ricky Rabbit to scare alligators. Ricky scared 16 alligators the first day and 18 alligators the next day. How many alligators did he scare in all?

$$\begin{array}{r} 16 \\ +18 \\ \hline 34 \end{array}$$

5. Waldo's dancing elephants did 16 shows on Monday and 19 shows on Tuesday. How many shows did they do altogether?

$$\begin{array}{r} 16 \\ +19 \\ \hline 35 \end{array}$$

6. An elephant named Butterfly learned 15 dances in the morning and 6 dances in the afternoon. How many dances did she learn in all?

$$\begin{array}{r} 15 \\ +6 \\ \hline 21 \end{array}$$

Page 91

Skill: Subtraction with Regrouping

The Busy Tooth Fairy

Name ________________

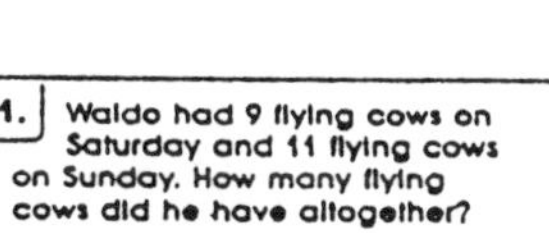

1. Adam Aardvark had 20 teeth. He lost 11 teeth. How many teeth did he have left?

$$\begin{array}{r} {}^{1}\not{2}{}^{1}0 \\ -11 \\ \hline 9 \end{array}$$

2. Candy Kitten had 26 teeth. She now has 17 teeth. How many teeth did the busy tooth fairy pay for?

$$\begin{array}{r} {}^{1}\not{2}{}^{1}6 \\ -17 \\ \hline 9 \end{array}$$

3. Harry Hare used to have 28 teeth. He still has 19 teeth. How many teeth did the tooth fairy pay for?

$$\begin{array}{r} {}^{1}\not{2}{}^{1}8 \\ -19 \\ \hline 9 \end{array}$$

4. Kenny Kangaroo had 34 teeth. 15 teeth fell out. How many were left?

$$\begin{array}{r} {}^{2}\not{3}{}^{1}4 \\ -15 \\ \hline 19 \end{array}$$

5. Peter Possum used to have 50 teeth. He lost 23 of them when he fell off the monkey bars. How many were left?

$$\begin{array}{r} {}^{4}\not{5}{}^{1}0 \\ -23 \\ \hline 27 \end{array}$$

6. Sam Seal lost 19 teeth. He used to have 34 teeth. How many teeth does he have now?

$$\begin{array}{r} {}^{2}\not{3}{}^{1}4 \\ -19 \\ \hline 15 \end{array}$$

7. Henry Hippo had 36 teeth. He lost 17 teeth. How many teeth does he have left?

$$\begin{array}{r} {}^{2}\not{3}{}^{1}6 \\ -17 \\ \hline 19 \end{array}$$

Page 92

Answer Key

Smelly Belly's Picnic

Skill: Addition & Subtraction with Regrouping

Name ______

1. Smelly Belly invited her friends to a picnic. She brought 17 bees, and her friend Stinky brought 14 bees. What was the total?
   17 + 14 = 31

2. Stinky brought 17 eggs to the picnic. Smelly brought 14 eggs. How many were there altogether?
   17 + 14 = 31

3. Smelly found 24 snails for the picnic. Only 18 were eaten. How many were left?
   24 − 18 = 6

4. Smelly loves to eat wasps. She brought 54 fried wasps. She ate 27 of them herself. How many wasps were left?
   54 − 27 = 27

5. Stinky came to the picnic with 43 crickets. Smelly came with 18 crickets. What was the total?
   43 + 18 = 61

6. Smelly Belly came to the picnic with 45 wild roots. The other skunks at the picnic ate 26 of them. How many roots were left?
   45 − 26 = 19

7. Stinky's favorite food is spiders. There were 32 at the picnic. He ate 29 spiders. How many were left?
   32 − 29 = 3

Page 93

Lizzy the Lizard — Bug Collector

Skill: Addition & Subtraction with Regrouping

Name ______

Lizzy the Lizard loves to collect bugs. She has a collection of beetles, crickets and flies. She would have more, but sometimes she forgets and eats part of her collection.

1. Lizzy caught 18 beetles and 21 flies. How many more flies did she catch?
   21 − 18 = 3

2. Lizzy found 19 bugs in a garden. Her friend Dizzy found 17. How many did they find in all?
   19 + 17 = 36

3. Lizzy had 31 flies on Sunday. She forgot and ate 12 of them. How many were left?
   31 − 12 = 19

4. Lizzy chased 34 butterflies. She caught 25 of them. How many got away?
   34 − 25 = 9

5. Lizzy found 12 beetles on Monday and 19 beetles on Tuesday. How many did she find in all?
   12 + 19 = 31

6. Lizzy caught 15 moths and 17 stinkbugs. How many insects did she catch altogether?
   15 + 17 = 32

Page 94

Worm Burgers

Skill: Working with Money (addition)

Name ______

Worm Burgers is a new fast-food place opened up near your school by someone who thinks that worms are tasty food. Would you like to go there for lunch some day?

1. You buy a Worm Burger for 50¢ and a Worm Cola for 10¢. How much do you spend?
   50¢ + 10¢ = 60¢

2. Your teacher buys a Worm Burger for 50¢ and Worm Fries for 25¢. How much does she spend in all?
   50¢ + 25¢ = 75¢

3. Your best friend buys a Worm Cola for 10¢ and Worm Fries for 25¢. How much does he spend?
   10¢ + 25¢ = 35¢

4. Anne buys a Worm Burger for 50¢. Alice buys Worm Fries for 25¢. Abby buys a Super Worm Burger for 70¢. Your teacher pays for it all. How much does it cost her?
   50¢ + 25¢ + 70¢ = 145¢ or $1.45

5. Your teacher is hungry. She buys a Worm Burger for 50¢ and a Super Worm Burger for 70¢. How much does she spend?
   50¢ + 70¢ = 120¢ or $1.20

6. You buy a Worm Cola for 10¢. Your best friend buys a Worm Cola for 10¢. Your teacher buys a Worm Cola, too. How much does it cost for all three Worm Colas?
   10¢ + 10¢ + 10¢ = 30¢

Page 95

Worm Burger Sale

Skill: Working with Money (subtraction, some regrouping)

Name ______

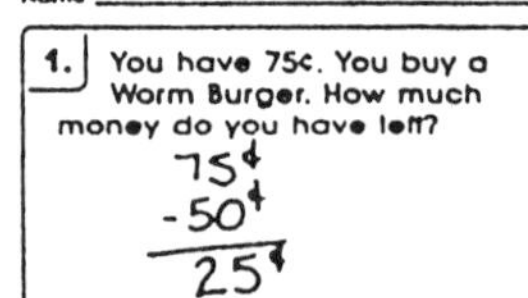

1. You have 75¢. You buy a Worm Burger. How much money do you have left?
   75¢ − 50¢ = 25¢

2. A Worm Berry Shake costs 85¢. You have 95¢. How much change do you get?
   95¢ − 85¢ = 10¢

3. Your teacher bought Worm Fries to go with her lunch. How much change did she get back 50¢?
   50¢ − 25¢ = 25¢

4. A Worm Berry Shake costs 85¢. Worm Fries cost 25¢. How much more is a Worm Berry Shake?
   85¢ − 25¢ = 60¢

5. A Super Worm Burger costs 70¢. You have only 45¢. How much more money do you need?
   70¢ − 45¢ = 25¢

6. Al bought his lunch at **Worm Burgers.** He spent 95¢. Don spent 40¢ on his lunch. How much more did Al spend?
   95¢ − 40 = 55¢

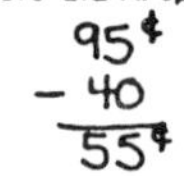

7. You get 80¢ from your mother to eat at **Worm Burgers.** You spent 60¢. How much do you have left?
   80¢ − 60¢ = 20¢

Page 96

Answer Key

Pat's Pet Shop

Skill: Working with Money (addition & subtraction, some regrouping)

Name ______________________

1. You went to Pat's Pet Shop to buy birdseed which cost 59¢ and rabbit food which cost 39¢. What was the sum?
   59¢ + 39¢ = 98¢

2. You bought a stuffed cat for your pet mouse to play with. It cost 44¢. You gave Pat 50¢. How much change did you get?
   50¢ − 44¢ = 6¢

3. Your friend bought a bottle of fish food for 39¢ and a can of cat food for 29¢. What was the total?
   39¢ + 29¢ = 68¢

4. Your teacher bought a bag of seed for her pet bird. It cost 39¢. She gave Pat 75¢. How much change did she get?
   75¢ − 39¢ = 36¢

5. Your best friend needed a play mouse for his cat. It cost 67¢. He gave Pat 75¢. How much change did he get?
   75¢ − 67¢ = 8¢

6. Pat sold your teacher a pet mouse for 65¢ and a pet lizard for 34¢. What was the total?
   65¢ + 34¢ = 99¢

7. The large box of goldfish food costs 98¢. The small size costs 49¢. What is the difference?
   98¢ − 49¢ = 49¢

Page 97

Bob Z. Cat

Skill: Working with Money (addition and subtraction)

Name ______________________

Bob Z. Cat is a very odd cat. He is afraid of mice. He loves dogs, and he can't wait to take a bath.

1. Bob Cat has a bottle of bubble bath which cost a quarter and smelly soap which cost a dime. What is the total?
   25¢ + 10¢ = 35¢

2. Bob Cat has a toy mouse which cost 43¢ and a toy rat which cost 34¢. What is the sum?
   43¢ + 34¢ = 77¢

3. Bob Cat used to have 86¢. He spent 54¢ on a ball of string. How much does he have left?
   86¢ − 54¢ = 32¢

4. Bob Cat had 89¢ in his piggy bank. He spent 42¢. How much was left?
   89¢ − 42¢ = 47¢

5. Bob Cat was given catnip which cost 56¢ and a toy mouse which cost 45¢. What was the sum?
   56¢ + 45¢ = 101¢ or $1.01

6. Bob Z. Cat could have the large can of cat food which costs 69¢ or the small can which costs 51¢. What is the difference?
   69¢ − 51¢ = 18¢

Page 98

Animal Trivia

Skill: Metric Measurement (centimeters)

Name ______________________

1. The house mouse has a tail 10 cm long. The pack rat has a tail 19 cm long. How much longer is the rat's tail?
   19 cm − 10 cm = 9 cm

2. A chipmunk is 24 cm long. A gopher is 19 cm long. How much longer is the chipmunk?
   24 cm − 19 cm = 5 cm

3. The gray squirrel has a tail 24 cm long. The red squirrel has a tail 16 cm long. How much longer is the tail of the gray squirrel?
   24 cm − 16 cm = 8 cm

4. The spotted bat has a body that is 11 cm long. The tail is 5 cm long. How long is the bat from the tip of his nose to the tip of his tail?
   11 cm + 5 cm = 16 cm

5. The jackrabbit has a back foot that is 17 cm long. The back foot of the cottontail rabbit is 10 cm long. How much longer is the back foot of the jackrabbit?
   17 cm − 10 cm = 7 cm

6. The back foot of a polar bear is 33 cm long. The back foot of a black bear is 37 cm long. How much longer is the black bear's foot?
   37 cm − 33 cm = 4 cm

7. The body of a pocket mouse is 20 cm long. His tail is 14 cm long. How long is the pocket mouse from the tip of his nose to the tip of his tail?
   20 cm + 14 cm = 34 cm

Page 99

Lizzy the Lizard

Skill: 1-digit Multiplication

Name ______________________

1. Lizzy the Lizard ate 3 beetles for breakfast. She ate 3 times as many for lunch. How many did she eat for lunch?
   $3 \times 3 = 9$

2. Lizzy ate 4 plates of beetles. Each plate had 3 beetles. How many beetles did she eat?
   $4 \times 3 = 12$

3. Lizzy the Lizard caught 5 bugs. Her friend Dizzy caught 3 times as many bugs. How many bugs did Dizzy catch?
   $5 \times 3 = 15$

4. Lizzy had 3 piles of bugs. Each pile had 4 bugs. How many bugs did she have?
   $3 \times 4 = 12$

5. Lizzy had 2 crickets for a snack. She had 4 snacks in a day. How many crickets did she eat?
   $2 \times 4 = 8$

6. Lizzy caught 4 ants. Her friend Dizzy caught 2 times as many ants. How many ants did Dizzy catch?
   $4 \times 2 = 8$

7. Lizzy counted 6 beetles in an hour. She counted 2 times as many in the next hour. How many did she count in the next hour?
   $6 \times 2 = 12$

Page 100

Answer Key

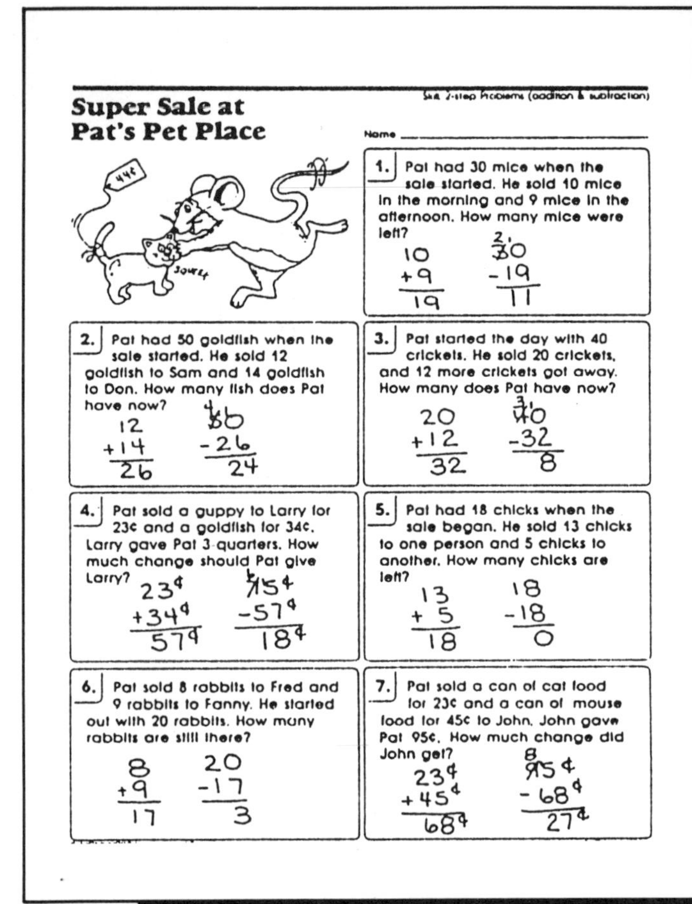

Super Sale at Pat's Pet Place

Skill: 2-step Problems (addition & subtraction)

Name ____________________

1. Pat had 30 mice when the sale started. He sold 10 mice in the morning and 9 mice in the afternoon. How many mice were left?
   10 + 9 = 19; 30 − 19 = 11

2. Pat had 50 goldfish when the sale started. He sold 12 goldfish to Sam and 14 goldfish to Don. How many fish does Pat have now?
   12 + 14 = 26; 50 − 26 = 24

3. Pat started the day with 40 crickets. He sold 20 crickets, and 12 more crickets got away. How many does Pat have now?
   20 + 12 = 32; 40 − 32 = 8

4. Pat sold a guppy to Larry for 23¢ and a goldfish for 34¢. Larry gave Pat 3 quarters. How much change should Pat give Larry?
   23¢ + 34¢ = 57¢; 75¢ − 57¢ = 18¢

5. Pat had 18 chicks when the sale began. He sold 13 chicks to one person and 5 chicks to another. How many chicks are left?
   13 + 5 = 18; 18 − 18 = 0

6. Pat sold 8 rabbits to Fred and 9 rabbits to Fanny. He started out with 20 rabbits. How many rabbits are still there?
   8 + 9 = 17; 20 − 17 = 3

7. Pat sold a can of cat food for 23¢ and a can of mouse food for 45¢ to John. John gave Pat 95¢. How much change did John get?
   23¢ + 45¢ = 68¢; 95¢ − 68¢ = 27¢

Page 101

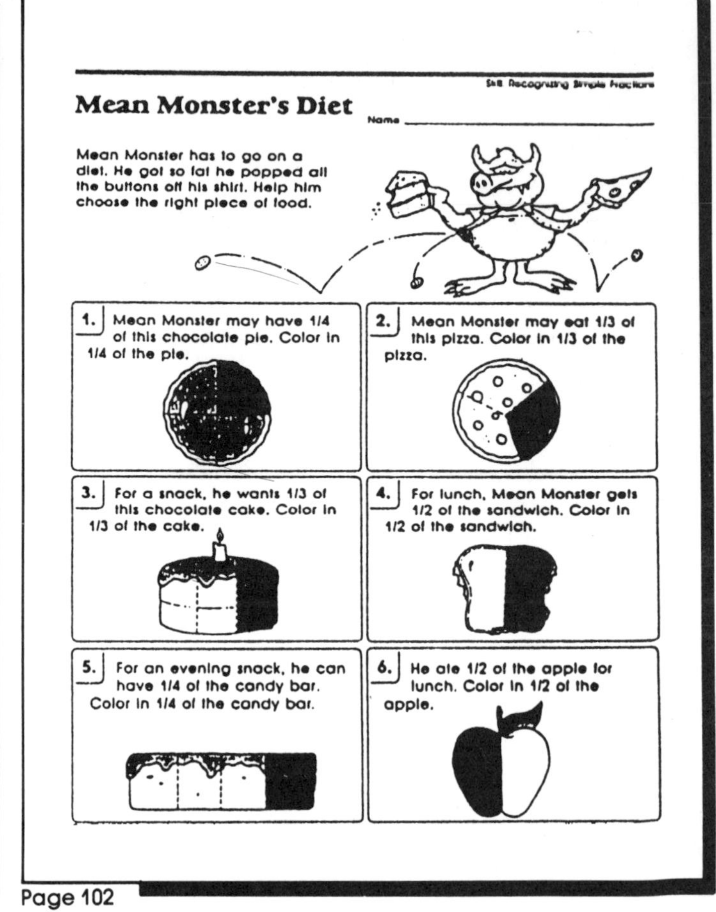

Mean Monster's Diet

Skill: Recognizing Simple Fractions

Name ____________________

Mean Monster has to go on a diet. He got so fat he popped all the buttons off his shirt. Help him choose the right piece of food.

1. Mean Monster may have 1/4 of this chocolate pie. Color in 1/4 of the pie.

2. Mean Monster may eat 1/3 of this pizza. Color in 1/3 of the pizza.

3. For a snack, he wants 1/3 of this chocolate cake. Color in 1/3 of the cake.

4. For lunch, Mean Monster gets 1/2 of the sandwich. Color in 1/2 of the sandwich.

5. For an evening snack, he can have 1/4 of the candy bar. Color in 1/4 of the candy bar.

6. He ate 1/2 of the apple for lunch. Color in 1/2 of the apple.

Page 102

About the book . . .

This book contains activity pages on such math topics as time, money, measurement and graphing, plus story problems with humorous twists and logical approaches to help students solve each one.

Credits . . .
Authors: Jan Kennedy, Paula Corbett, James E. Davidson, Ph.D., Robert W. Smith
Editors: Jackie Servis and Rhonda DeWaard
Artists: Ann Dyer, Pat Bakken, Karen Caminata, Jim Price, Carol Tiernon, Marj Waldschmidt
Production: Ann Dyer and Kurt Kemperman
Cover Photo: Dan Van Duinen